# Design, Production and Characterization of Peptide Antibodies

# Design, Production and Characterization of Peptide Antibodies

Editors

**Nicole Hartwig Trier**
**Gunnar Houen**

MDPI • Basel • Beijing • Wuhan • Barcelona • Belgrade • Manchester • Tokyo • Cluj • Tianjin

*Editors*
Nicole Hartwig Trier
Department of Neurology,
Rigshospitalet Glostrup
Glostrup
Denmark

Gunnar Houen
Department of Neurology,
Rigshospitalet Glostrup
Glostrup
Denmark

*Editorial Office*
MDPI
St. Alban-Anlage 66
4052 Basel, Switzerland

This is a reprint of articles from the Special Issue published online in the open access journal *Antibodies* (ISSN 2073-4468) (available at: https://www.mdpi.com/journal/antibodies/special_issues/Design_Production_Characterization_Peptide_Antibodies).

For citation purposes, cite each article independently as indicated on the article page online and as indicated below:

LastName, A.A.; LastName, B.B.; LastName, C.C. Article Title. *Journal Name* **Year**, *Volume Number*, Page Range.

**ISBN 978-3-0365-7690-9 (Hbk)**
**ISBN 978-3-0365-7691-6 (PDF)**

# Contents

# About the Editors

**Nicole Hartwig Trier**

Nicole Hartwig Trier (MSc, PhD) was educated in Biochemistry and Molecular Biology at the University of Southern Denmark and Copenhagen University. Her focus areas are immunology and protein chemistry. She has published more than 60 scientific papers and book chapters. Her current research areas are autoimmune diseases and antibody structure and function in collaboration with Gunnar Houen.

**Gunnar Houen**

Gunnar Houen (MSc, PhD, DSc) was educated in Biochemistry with a special focus on Protein Chemistry and Molecular Biology. He has published more than 200 scientific papers and educated more than fifty MScs and PhDs. His current research interests are autoimmune diseases and antibody structure and function.

*antibodies*

MDPI

*Editorial*

# Design, Production, Characterization, and Use of Peptide Antibodies

**Nicole H. Trier [1,*] and Gunnar Houen [1,2,*]**

1 Department of Neurology, Rigshospitalet Glostrup, Valdemar Hansens Vej 13, 2600 Glostrup, Denmark
2 Department of Biochemistry and Molecular Biology, University of Southern Denmark, Campusvej 55, 5223 Odense, Denmark
* Correspondence: nicole.hartwig.trier@gmail.com (N.H.T.); gunnar.houen@regionh.dk (G.H.)

**Citation:** Trier, N.H.; Houen, G. Design, Production, Characterization, and Use of Peptide Antibodies. *Antibodies* **2023**, *12*, 6. https://doi.org/10.3390/antib12010006

Received: 14 July 2022
Revised: 10 October 2022
Accepted: 26 December 2022
Published: 13 January 2023



Antibodies are key reagents in diagnostics, therapeutics, and experimental biology, capable of detecting numerous targets. The immunization of animals (rabbits, mice, etc.) with antigens mixed with an adjuvant and the induction of antibodies is a well-established process. This procedure for the generation of monoclonal antibodies was originally described by Köhler and Milstein, and is well-known and has been thoroughly tested [1–5]. Proteins were originally the focus for antibody production; however, in cases where the native protein is not available or modified targets are of interest, synthetic peptides coupled with an immunogenic carrier can be used as antigens [2–7].

The choice of suitable peptides is crucial for the generation of a good peptide antibody. The peptides used are typically 10–20 amino acids long and can be designed to represent many different targets, e.g., post-translational modifications, terminal ends, areas of high conservation, turns, loops, α-helices, etc. [2,3,5]. Based on this, peptide antibodies have many applications, especially because they can easily be produced against multiple targets [2–11]. For example, mutation-specific peptide antibodies have become important research and diagnostic tools in malignant and premalignant conditions [5,9,11].

In addition to peptide antibodies obtained by the immunization of animals, they can be produced using mammalian expression systems, which facilitate native antibody folding and post-translational modifications [12]. This approach typically employs the co-transfection of heavy chain (HC) and light chain (LC) genes on separate plasmids [12,13], although the use of a bidirectional vector encoding both genes on a single plasmid may be applied, as recently described by Carrara et al. [14]. Approaches such as these may enable efficient small-scale antibody production by transient cell transfection. Another well-established method is phage display, based on the genetic engineering of bacteriophages and repeated rounds of antigen-guided selection and phage propagation [15,16]. This approach was recently applied by Anderson et al. to generate antibodies specific for a Lassa virus protein [16]. Crucial in this approach is the selection of phage-displayed antibody libraries using synthetic peptides as targets, resulting in the generation of peptide-specific antibodies, commonly referred to as phage-specific peptide antibodies [15]. Using phase display, the in vitro production of antibodies to virtually any target is feasible.

Antibodies are typically expressed in a basic format with two HCs and two LCs ($HC_2LC_2$) linked via disulfide bonds, although some antibodies are occasionally prone to dimerization, or even oligomerization, as recently described by Mieczkowski et al. [17,18]. However, other antibody formats exist [4,13,15], for example, Anderson et al. recently described the production of a single-domain antibody consisting of a monomeric variable antibody domain which may have diagnostic and therapeutic potential [16]. Another example includes a recently published model developed by Ramasubramanian et al., who described a bivalent IgG where two different variable HC domains with distinct binding specificities were grafted onto the first constant HC and constant LC domains, resulting in an antibody with dual specificity [13]. Antibodies composed only of two HCs are commonly found in sharks and camelids, and were recently produced by Mieczkowski et al., showing good physiochemical properties and stability [17].

Independent of antibody design and production method, relevant antibodies should be identified by examining binding to the respective targets in the assay, where they are intended to be used, in order to avoid assay restrictions [2,3,19,20]. For example, peptide antibodies do not necessarily recognize native structures or do not necessarily detect the final target in all immunoassay formats [3,19,20]. Thus, it is essential to test the produced antibodies in a variety of formats, and ideally in the format of the intended use.

Following selection, (peptide) antibodies should be characterized in terms of reactivity, specificity, and cross-reactivity, e.g., by titration assays and competition studies, where the relevant target is used as an inhibitor [2,3,5]. Moreover, biochemical characteristics such as isotype/subtype, solubility, stability, and binding characteristics should be determined [3]. Antibodies may be characterized systematically in relation to the antigenic epitope by various mapping techniques [2,3,5,21]. For antibodies recognizing a conformational epitope, this may involve mass spectrometry, nuclear magnetic resonance, X-ray crystallography, and surface plasmon resonance, as described by Ramasubramanian et al., Anderson et al., and Mieczkowski et al.; however, these are time-consuming approaches [13,16,18]. In contrast, if the epitope of the respective antibody is continuous, which typically applies to peptide antibodies, but also to some antibodies generated for full-length proteins or protein fragments [5,21], antibodies are more easily characterized by epitope mapping or alternatively by analyzing antibody reactivity to peptides with sequence homology, as recently described by Bergmann et al. and Fanelli et al., respectively [21,22]. Here, amino acids essential for antibody binding are typically identified by Ala scanning and "functionality" scanning after the terminal "borders" crucial for antibody reactivity have been identified using systematically N- and C-terminally truncated peptides [5,21,23]. For the characterization of peptide antibodies, resin-bound peptides have been applied with success, because this approach is rapid and saves time compared with the use of free peptides [5,21,23]. The advantage of using resin-bound peptides is that they may be tested for binding activity directly on a solid support [5,21,23], where the peptide is connected to the resin through the -OH group of the C-terminal amino acid. This allows the peptide to fold into a conformation resembling the native conformation, whereas during the coating of free peptides on a solid surface, the peptide may change conformation, or crucial amino acid side chains necessary for the detection of antibody binding may be masked [3,19].

Peptide antibodies are used for research and diagnostics in many different immunoassays, e.g., immunohistochemistry, immunocytochemistry, immunoblotting, immunoprecipitation, and sandwich assays [5,8–11,15,24]. In the diagnostic field, peptide antibodies may aid in the diagnosis of diseases and infections [8,9,11]. For example, peptide antibodies have been used in immunohistochemistry for the detection of specific cancerous point mutations with great success, such as peptide antibodies specific for point mutations or deletions in B-raf, isocitrate dehydrogenase, and epidermal growth factor receptor, which are associated with melanoma, lung adenocarcinoma, and gliomas, respectively [9,11]. Moreover, a peptide antibody to the frame-shifted C-terminal of calreticulin was recently described; it showed promising results as a diagnostic candidate for calreticulin-associated myeloproliferative disorders [5]. Additionally, peptide antibodies have been used for detecting other tumors [10,11,15]. Moreover, peptide antibodies have been used for diagnosing neurodegenerative diseases, disorders of the immune system, infectious diseases, cardiovascular diseases, and other disorders, and to detect the presence of various viruses, bacteria, and parasites [8,11,25]. Despite obvious potential, the clinical therapeutic use of peptide antibodies still needs to be demonstrated. A promising example is peptide antibodies to gp120, which have been shown to neutralize HIV-1 infectivity in vitro [6]. Similar, a peptide antibody to procathepsin D has been proposed to inhibit breast cancer development, and treatment with a peptide antibody to apoB has been shown to reduce atherosclerosis in hypercholesterolemic mice, indicating that this antibody has therapeutic potential [26,27]. Finally, a peptide antibody to C5a, a protein fragment released from the cleavage of complement component C5 by the protease C5-convertase, has shown promising results in inhibiting sepsis [24]. Collectively, the abovementioned examples indicate that peptide antibodies have an unresolved potential in the therapeutic field.

The seven articles featured in this Special Issue, addressing the design, production, characterization, and use of (peptide) antibodies, are of considerable interest to scientists investigating approaches related to this topic.

**Conflicts of Interest:** The authors declare no conflict of interest.

## References


1. Kohler, G.; Howe, S.C.; Milstein, C. Fusion between immunoglobulin-secreting and nonsecreting myeloma cell lines. *Eur. J. Immunol.* **1976**, *6*, 292–295. [CrossRef] [PubMed]
2. Trier, N.H.; Hansen, P.R.; Houen, G. Production and characterization of peptide antibodies. *Methods* **2012**, *56*, 136–144. [CrossRef] [PubMed]
3. Trier, N.H.; Mortensen, A.; Schiolborg, A.; Friis, T. Production and screening of monoclonal peptide antibodies. *Methods Mol. Biol.* **2015**, *1348*, 109–126. [PubMed]
4. Houen, G. *Peptide Antibodies: Methods and Protocols*; Humana Press: Totowa, NJ, USA, 2016.
5. Mughal, F.P.; Bergmann, A.C.; Huynh, H.U.B.; Jørgensen, S.H.; Mansha, I.; Kesmez, M.; Schürch, P.M.; Theocharides, A.P.A.; Hansen, P.R.; Friis, T.; et al. Production and Characterization of Peptide Antibodies to the C-Terminal of Frameshifted Calreticulin Associated with Myeloproliferative Diseases. *Int. J. Mol. Sci.* **2022**, *23*, 6803. [CrossRef] [PubMed]
6. Prieto, I.; Hervás-Stubbs, S.; García-Granero, M.; Berasain, C.; Riezu-Boj, J.I.; Lasarte, J.J.; Sarobe, P.; Prieto, J.; Borrás-Cuesta, F. Simple strategy to induce antibodies of distinct specificity: Application to the mapping of gp120 and inhibition of HIV-1 infectivity. *Eur. J. Immunol.* **1995**, *25*, 877–883. [CrossRef]
7. Park, B.K.; Lee, S.I.; Bae, J.Y.; Park, M.S.; Lee, Y.; Kwon, H.J. Production of a monoclonal antibody targeting the M protein of MERS-CoV for detection of MERS-CoV using a synthetic peptide epitope formulated with a CpG-DNA-liposome complex. *Int. J. Pept. Res. Ther.* **2019**, *25*, 819–826. [CrossRef]
8. Bhullar, S.S.; Chandak, N.H.; Baheti, N.N.; Purohit, H.J.; Taori, G.M.; Daginawala, H.F.; Kashyap, R.S. Diagnosis of herpes simplex encephalitis by elisa using antipeptide antibodies against type-common epitopes of glycoprotein b of herpes simplex viruses. *J. Immunoass. Immunochem.* **2016**, *37*, 217–227. [CrossRef]
9. Chen, Q.; Xia, C.; Deng, Y.; Wang, M.; Luo, P.; Wu, C.; Yue, J.; Fang, N.; Wang, M.; Wei, S. Immunohistochemistry as a quick screening method for clinical detection of braf(v600e) mutation in melanoma patients. *Tumor Biol.* **2014**, *35*, 5727–5733. [CrossRef]
10. Field, S.; Uyttenhove, C.; Stroobant, V.; Cheou, P.; Donckers, D.; Coutelier, J.P.; Simpson, P.T.; Cummings, M.C.; Saunus, J.M.; Reid, L.E.; et al. Novel highly specific anti-periostin antibodies uncover the functional importance of the fascilin 1-1 domain and highlight preferential expression of periostin in aggressive breast cancer. *Int. J. Cancer* **2016**, *138*, 1959–1970. [CrossRef]
11. Trier, N.H.; Houen, G. Peptide Antibodies in Clinical Laboratory Diagnostics. *Adv. Clin. Chem.* **2017**, *81*, 43–96.
12. Schlatter, S.; Stansfield, S.H.; Dinnis, D.M.; Racher, A.J.; Birch, J.R.; James, D.C. On the optimal ratio of heavy to light chain genes for efficient recombinant antibody production by CHO cells. *Biotechnol. Prog.* **2005**, *21*, 122–133. [CrossRef]
13. Ramasubramanian, A.; Tennyson, R.; Magnay, M.; Kathuria, S.; Travaline, T.; Jain, A.; Lord, D.M.; Salemi, M.; Sullican, C.; Magnay, T.; et al. Bringing the heavy Chain to Light: Creating a symmetric, bivalent IgG-like bispecific. *Antibodies* **2020**, *9*, 62. [CrossRef]
14. Carrara, S.C.; Fiebig, D.; Bogen, J.P.; Grzeschik, J.; Hock, B.; Kolmar, H. Recombinant antibody production using a dual-promoter single plasmid system. *Antibodies* **2021**, *10*, 18. [CrossRef]
15. van Beijnum, J.R.; Moerkerk, P.T.M.; Gerbers, A.J.; de Bruïne, A.P.; Arends, J.-W.; Hoogenboom, H.R.; Hufton, S.E. Target validation for genomics using peptide-specific phage antibodies: A study of five gene products overexpressed in colorectal cancer. *Int. J. Cancer* **2002**, *101*, 118–127. [CrossRef]
16. Anderson, G.P.; Liu, J.L.; Shriver-Lake, L.C.; Goldman, E.R. Selection and characterization of single-domain antibodies for detection of lassa nucleoprotein. *Antibodies* **2020**, *9*, 71. [CrossRef]
17. Mieczkowski, C.; Bahmanjah, S.; Yu, Y.; Baker, J.; Raghunathan, G.; Tomazela, D.; Hsieh, M.; McCoy, M.; Strickland, C.; Fayadat-Dilman, L. Crystal structure and characterization of human heavy-chain only antibodies reveals a novel, stable dimeric structure similar to monoclonal antibodies. *Antibodies* **2020**, *9*, 66. [CrossRef]
18. Mieczkowski, C.; Chenh, A.; Fischmann, T.; Hsieh, M.; Baker, J.; Uchida, M.; Raghunathan, G.; Strickland, C.; Fayadat-Dilman, L. Characterization and modeling of reversible antibody self-association provide insights into behavior, prediction and correction. *Antibodies* **2021**, *10*, 8. [CrossRef]
19. Holm, B.E.; Bergmann, A.C.; Hansen, P.R.; Koch, C.; Houen, G.; Trier, N.H. Antibodies with specificity for native and denatured forms of ovalbumin differ in reactivity between enzyme-linked immunosorbent assays. *APMIS* **2015**, *123*, 136–145. [CrossRef]
20. Nelson, P.N.; Fletcher, S.M.; MacDonald, D.; Goodall, D.M.; Jefferis, R. Assay restriction profiles of three monoclonal antibodies recognizing G3m(u)allotype. Development of an allotype specific assay. *J. Immunol. Methods* **1991**, *138*, 57–64. [CrossRef]
21. Bergmann, A.C.; Kyllesbech, C.; Slibinskas, R.; Ciplys, E.; Højrup, P.; Trier, N.H.; Houen, P. Epitope mapping of monoclonal antibodies to calreticulin reveals that charged amino acids are essential for antibody binding. *Antibodies* **2021**, *10*, 31. [CrossRef]

22. Fanelli, I.; Rovero, P.; Hansen, P.R.; Frederiksen, J.; Houen, G.; Trier, N.H. Specificity of anti-citrullinated protein antibodies to citrullinated α-enolase peptides as a function of epitope structure and composition. *Antibodies* **2021**, *10*, 27. [CrossRef] [PubMed]
23. Trier, N.H. Characterization of peptide antibodies by epitope mapping using resin-bound and soluble peptides. *Methods Mol. Biol.* **2015**, *1348*, 229–239. [PubMed]
24. Huber-Lang, M.S.; Sarma, J.V.; McGuire, S.R.; Lu, K.T.; Guo, R.F.; Padgaonkar, V.A.; Younkin, E.M.; Laudes, I.J.; Riedemann, N.C.; Younger, J.G.; et al. Protective effects of anti-C5a peptide antibodies in experimental sepsis. *FASEB J.* **2001**, *15*, 568–570. [CrossRef] [PubMed]
25. Stock, N.K.; Escadafal, C.; Achazi, K.; Cissé, M.; Niedrig, M. Development and characterization of polyclonal peptide antibodies for the detection of Yellow fever virus proteins. *Methods* **2015**, *222*, 110–116. [CrossRef]
26. Vetvicka, V.; Vetvickova, J.; Fusek, M. Anti-human procathepsin D activation peptide antibodies inhibit breast cancer development. *Breast Cancer Res. Treatment* **1999**, *57*, 261–269. [CrossRef]
27. Gonçalves, I.; Nitulescu, M.; Ares, M.P.S.; Fredrikson, G.N.; Jansson, B.; Li, Z.-C.; Nilsson, J. Identification of the target for therapeutic recombinant anti-apoB-100 peptide antibodies in human atherosclerotic lesions. *Atherosclerosis* **2009**, *205*, 96–100. [CrossRef]

antibodies

Article

# Epitope Mapping of Monoclonal Antibodies to Calreticulin Reveals That Charged Amino Acids Are Essential for Antibody Binding

**Ann Christina Bergmann [1], Cecilie Kyllesbech [2], Rimantas Slibinskas [3], Evaldas Ciplys [3], Peter Højrup [4], Nicole Hartwig Trier [2,*] and Gunnar Houen [2,4,*]**

1 Department of Autoimmunology, Statens Serum Institut, 2300 Copenhagen, Denmark; ann_bergmann@hotmail.com
2 Department of Neurology, Rigshospitalet Glostrup, 2600 Glostrup, Denmark; cekr16@student.sdu.dk
3 Institute of Biotechnology, University of Vilnius, 01513 Vilnius, Lithuania; rimantas.slibinskas@bti.vu.lt (R.S.); evaldas.ciplys@bti.vu.lt (E.C.)
4 Department of Biochemistry and Molecular Biology, University of Southern Denmark, 5230 Odense, Denmark; php@bmb.sdu.dk
* Correspondence: nicole.hartwig.trier@regionh.dk (N.H.T.); gunnar.houen@regionh.dk (G.H.)

**Abstract:** Calreticulin is a chaperone protein, which is associated with myeloproliferative diseases. In this study, we used resin-bound peptides to characterize two monoclonal antibodies (mAbs) directed to calreticulin, mAb FMC 75 and mAb 16, which both have significantly contributed to understanding the biological function of calreticulin. The antigenicity of the resin-bound peptides was determined by modified enzyme-linked immunosorbent assay. Specific binding was determined to an 8-mer epitope located in the N-terminal (amino acids 34–41) and to a 12-mer peptide located in the C-terminal (amino acids 362–373). Using truncated peptides, the epitopes were identified as TSRWIESK and DEEQRLKEEED for mAb FMC 75 and mAb 16, respectively, where, especially the charged amino acids, were found to have a central role for a stable binding. Further studies indicated that the epitope of mAb FMC 75 is assessable in the oligomeric structure of calreticulin, making this epitope a potential therapeutic target.

**Keywords:** antibody; calreticulin; epitope; enzyme-linked immunosorbent assay

**Citation:** Bergmann, A.C.; Kyllesbech, C.; Slibinskas, R.; Ciplys, E.; Højrup, P.; Trier, N.H.; Houen, G. Epitope Mapping of Monoclonal Antibodies to Calreticulin Reveals That Charged Amino Acids Are Essential for Antibody Binding. *Antibodies* **2021**, *10*, 31. https://doi.org/10.3390/antib10030031

Academic Editor: Tianlei Ying

Received: 20 May 2021
Accepted: 30 July 2021
Published: 4 August 2021



## 1. Introduction

Calreticulin (CRT) is a molecular chaperone protein mainly located in the endoplasmic reticulum (ER), where it also functions as a $Ca^{2+}$ storage protein [1,2]. CRT has been assigned multiple other biological and immunological functions in addition to the above mentioned, including a role as a surface opsonin ("eat me signal") in connection with surface translocation during so-called "immunogenic cell death" [1,3–8]. Moreover, characteristic mutations in the C-terminus of CRT, mainly insertions and deletions (INDELS), cause a frameshift (fs) change to a common multibasic sequence (CRTfs), which has been found to be involved in several myeloproliferative neoplasms (MPNs) [9–13].

The structure of CRT is unusual. The N-terminal 200 amino acids (aas) form the major part of a β sandwich, from which a proline-rich hairpin of approximately 100 aas protrudes. This domain continues into a C-terminal part of another 100 aas, which contributes two strands to the β sandwich and an α-helix, which together with the N-terminal part constitutes a globular core. The rest of the C-terminus is strongly acidic and can bind multiple $Ca^{2+}$, thus stabilizing the molecule, and it ends in a KDEL ER retention signal [14–18]. Both wild-type (wt) CRT and CRTfs can form oligomers, a property that is important for the properties of the molecules [19–24].

Antibodies (Abs) in general are crucial reagents in the study of the biological and immunological functions of proteins and other macromolecules, and monoclonal Abs (mAbs) and peptide Abs have revolutionized molecular and cellular biology [25–28]. The importance of Abs as reagents applies in particular to CRT and CRTfs, due to their unusual structures and many diverse functions. Several Abs have been described, and some are commercially available, but none have been adequately characterized in relation to interaction with specific parts of the chaperone, although some knowledge of their epitopes is available [29–31]. Here, we report the epitope mapping of two important mAbs to CRT, which have contributed to understanding the biological function of CRT [32–38].

Although several studies have been described using the two selected mAbs, the aas crucial for antigen (Ag) binding still remain to be characterized. Moreover, the study of theseAbs may contribute to determining the structure of the CRT protein, as only a part of CRT has been crystalized [16,18].

## 2. Materials and Methods

### 2.1. Materials

All synthetic peptides were purchased from Schäfer-N, Copenhagen, Denmark. CRT mAb FMC 75 (Cat.no. ADI-SPA-601) and mAb 16 (Cat.no. 612137) were purchased from Enzo Life Sciences and BD Transduction laboratories, respectively. MAb FMC 75 is a mouse IgG1 directed to full-length recombinant CRT, whereas mAb 16 is a mouse IgG1 directed to the C-terminal of CRT (aas 270–390). Both Abs were generated in mice using traditional immunization techniques for Ab production, where the mice were immunized with recombinant full-length CRT or a protein fragment. Tris-Tween-NaCl (TTN) (0.3 M NaCl, 20 mM Tris, 0.01% Tween 20, pH 7.5), carbonate buffer (0.05 M sodium carbonate, pH 9.6), alkaline phosphatase (AP) substrate buffer (1 M ethanolamine, 0.5 mM $MgCl_2$, pH 9.8), native sample buffer (Tris-HCL 0.5M pH 8.6, glycerol 99%, bromphenylblåt 0.1% *w/v*), mouse IgG1 mAbs directed to β-galactosidase and to BAM were from Statens Serum Institut (Copenhagen, Denmark). *Para*-nitrophenyl phosphate (*p*-NPP) substrate tablets, BCIP/NTB tablets, and goat anti-mouse IgG-AP were purchased from Sigma Aldrich (Steinheim, Germany). Recombinant CRT was produced using standard protocols and purified as previously described by Čiplys and colleagues [39]. Precision Plus Protein Standard was purchased from BIO-RAD. Native PAGE running buffer (pH 7.5), 4–20% Tris-glycine gels and, PVDF membranes were from Thermo–Fisher (Waltman, MA, USA).

### 2.2. Design of Synthetic Peptides

The human CRT sequence (Prot id: P27797), comprising 400 aas (without the N-terminal signal peptide), was used to generate overlapping peptides. The peptides were 20 aas long, each with an overlap of 10 aas (Table A1).

To determine essential terminal aas, N- and C-terminally truncated peptides were applied (Table 1), whereas alanine and functionally substituted peptides were used to determine essential aas crucial for Ab binding (Table 1). All peptides were synthesized and tested on the resin, which allows rapid epitope identification [40–42]. The peptides were synthesized on a TentaGel resin using standard solid-phase peptide synthesis.

**Table 1.** Overview of peptides used for antibody characterization.

| | *N*-Terminal Truncated Peptide | C-Terminal Truncated Peptides | Functionality-Substituted Peptides | Ala-Substituted Peptides |
|---|---|---|---|---|
| | | mAb FMC 75 | | |
| Template | GDGWTSRWIESKHKSD | | TSRWIESK | |
| | DGWTSRWIESKHKSD | GDGWTSRWIESKHKS | **S**SRWIESK | **A**SRWIESK |
| | GWTSRWIESKHKSD | GDGWTSRWIESKHK | T**T**RWIESK | T**A**RWIESK |
| | WTSRWIESKHKSD | GDGWTSRWIESKH | TS**K**WIESK | TS**A**WIESK |
| | TSRWIESKHKSD | GDGWTSRWIESK | TSR**F**IESK | TSR**A**IESK |
| | SRWIESKHKSD | GDGWTSRWIES | TSRW**L**ESK | TSRW**A**ESK |
| | RWIESKHKSD | GDGWTSRWIE | TSRWI**D**SK | TSRWI**A**SK |
| | WIESKHKSD | GDGWTSRWI | TSRWIE**T**K | TSRWIE**A**K |
| | IESKHKSD | GDGWTSRW | TSRWIES**R** | TSRWIES**A** |
| | ESKHKSD | GDGWTSR | | |
| | SKHKSD | GDGWTS | | |
| | KHKSD | GDGWT | | |
| | HKSD | GDGW | | |
| | | mAb 16 | | |
| Template | KDKQDEEQRLKEEEEDKKRK | | DEEQRLKEEEED | |
| | QDEEQRLKEEEEDKK | KQDEEQRLKEEEEDK | **E**EEQRLKEEEED | **A**EEQRLKEEEED |
| | DEEQRLKEEEEDKK | KQDEEQRLKEEEED | D**D**EQRLKEEEED | D**A**EQRLKEEEED |
| | EEQRLKEEEEDKK | KQDEEQRLKEEEE | DE**D**QRLKEEEED | DE**A**QRLKEEEED |
| | EQRLKEEEEDKK | KQDEEQRLKEEE | DEE**N**RLKEEEED | DEE**A**RLKEEEED |
| | QRLKEEEEDKK | KQDEEQRLKEE | DEEQ**K**LKEEEED | DEEQ**A**LKEEEED |
| | RLKEEEEDKK | KQDEEQRLKE | DEEQR**I**KEEEED | DEEQR**A**KEEEED |
| | LKEEEEDKK | KQDEEQRLK | DEEQRL**R**EEEED | DEEQRL**A**EEEED |
| | KEEEEDKK | KQDEEQRL | DEEQRLK**D**EEED | DEEQRLK**A**EEED |
| | EEEEDKK | KQDEEQR | DEEQRLKE**D**EED | DEEQRLKE**A**EED |
| | EEEDKK | KQDEEQ | DEEQRLKEE**D**ED | DEEQRLKEE**A**ED |
| | EEDKK | KQDEE | DEEQRLKEEE**D**D | DEEQRLKEEE**A**D |
| | EDKK | KQDE | DEEQRLKEEEE**E** | DEEQRLKEEEE**A** |

### 2.3. *Screening of Antibody Reactivity to Wild Type Calreticulin by Enzyme-Linked Immunosorbent Assay*

CRT (1 μg/mL) diluted in carbonate buffer was added to Polysorp microtiter plates (Thermo–Fisher, Waltman, MA, USA) and incubated overnight (ON) at room temperature (RT). Wells were rinsed with TTN (200 μL/well) and blocked in TTN buffer for 1 h (h) at RT on a shaking table. CRT mAb (1 μg/mL), diluted in TTN buffer, were added to the microtiter wells and incubated for 1 h at RT on a shaking table. Hereafter, the wells were rinsed with TTN (200 μL) for 1 min, and this was repeated 3 times. Next, AP-conjugated goat anti-mouse IgG (1 μg/mL) was added and incubated for 1 h at RT followed by washing as above. Bound Abs were quantified using 1 mg/mL *p*-NPP in AP substrate buffer, 100 μL per well. The absorbance was measured at 405 nm, with background subtraction at 650 nm on a ThermoMax Microtiter Plate Reader from Molecular Devices (San Jose, CA, USA). All samples were tested in duplicates and corrected for background noise.

### 2.4. *Screening of Resin-Bound Peptides by Modified Enzyme-Linked Immunosorbent Assay*

Resin-bound peptides (100 μg/mL) were added to a 96-well multiscreen filter plate (Millipore, Copenhagen, Denmark) and blocked in TTN for 15 min. CRT mAbs were diluted to a final concentration of 1 μg/mL in TTN buffer, added to the microtiter wells, and incubated for 1 h at RT. Hereafter the wells were rinsed with TTN (200 μL) for 1 min, and this was repeated three times. Next, AP-conjugated goat anti-mouse IgG diluted to 1 μg/mL was added and incubated for 1 h at RT followed by washing as above. Bound

Abs were quantified using 1 mg/mL *p*-NPP in AP substrate buffer, 100 µL pr well. After a satisfying color reaction, the buffer was transferred to a MaxiSorp Microtiter plate, and the absorbance was measured at 405 nm, with background subtraction at 650 nm on a ThermoMax Microtiter Plate Reader. All samples were tested in duplicates and corrected for background noise.

### 2.5. PAGE and Western Blotting

For Coomassie staining, recombinant CRT and CRT purified from human placenta were diluted in native sample buffer (1:2) and incubated for 10 min at 57 °C. Ten µL samples were loaded into wells of a 4–20% Tris-glycine gel and run for approximately 75 min at 150 V using native PAGE running buffer. Gels were stained ON at 4 °C with Coomassie Brilliant Blue and washed with Milli-Q water 5 times until bands were visualized.

For Western blotting, pretreated (57 °C for 10 min) recombinant CRT (1 mg/mL) (with and without 42 °C for three days) or CRT (1 mg/mL) purified from human placenta were loaded to a 4–20% Tris-glycine gel and run for approximately 75 min at 150 V using native PAGE running buffer, as described above.

After electrophoresis, gels were blotted onto a PVDF membrane using an iBlotter (Thermo–Fisher, Waltman, MA, USA). Next, the membranes were blocked ON in TTN buffer and mounted in a mini blotting device. Mab FMC 75 was diluted at 1:10,000 in TTN, added to each well, and incubated for 1 h at RT. Next, the membrane was washed for 3 × 5 min in TTN, whereafter AP-conjugated goat anti-mouse IgG, diluted at 1:2000, was added and incubated for 1 h on the membrane followed by washing 3 × 5 min in TTN. Finally, AP substrate (BCIP 0.5 mg/mL, NTB 0.3 mg/mL) was added and incubated for approximately 10 min. The reaction was stopped by washing the membrane in Milli-Q water, followed by drying it on filter paper.

## 3. Results

### 3.1. Reactivity of Monoclonal Antibodies to Wild Type Calreticulin

To analyze the specific reactivity of two selected mAbs to CRT, Ab reactivity to full-length CRT was determined by ELISA. A mAb generated in the same host but with irrelevant specificity was used as a negative control.

As seen in Figure 1, the two mAbs showed specific reactivity to the full-length recombinant wt CRT, as neither the negative control Ab nor the secondary Ab reacted with the CRT protein.

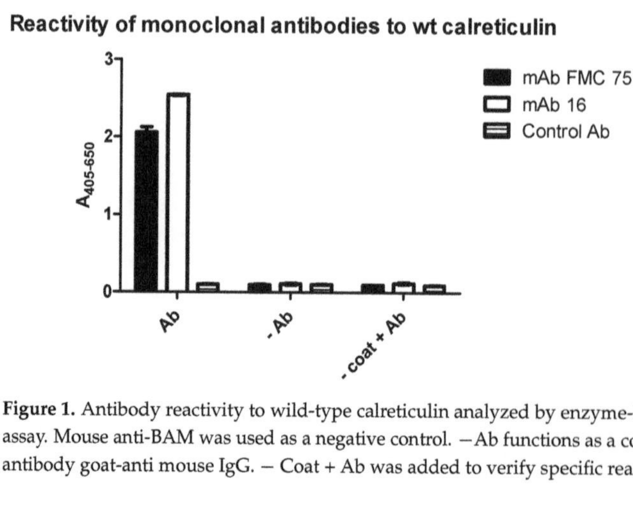


**Figure 1.** Antibody reactivity to wild-type calreticulin analyzed by enzyme-linked immunosorbent assay. Mouse anti-BAM was used as a negative control. −Ab functions as a control for the secondary antibody goat-anti mouse IgG. − Coat + Ab was added to verify specific reactivity.

### 3.2. Reactivity of Calreticulin Antibodies to Overlapping Synthetic Peptides

To identify the antigenic regions of the two mAbs, Ab reactivity to overlapping peptides covering the full-length CRT protein was determined by modified ELISA using resin-bound peptides.

As seen in Figure 2, mAb FMC 75 recognized peptide 2 in the N-terminus, whereas mAb 16 recognized a peptide in the C-terminus, corresponding to peptide 35. Peptide 2 corresponds to aas 28–47, whereas peptide 35 corresponds to aas 358–377 in the native CRT sequence. To verify that the interaction to peptide 2 and 35 was specific, a mAb of irrelevant specificity was used as a negative control. As seen, no reactivity to the overlapping CRT peptides was found, indicating that the interactions of mAb FMC 75 and mAb 16 to peptides 2 and 35, respectively, were specific.

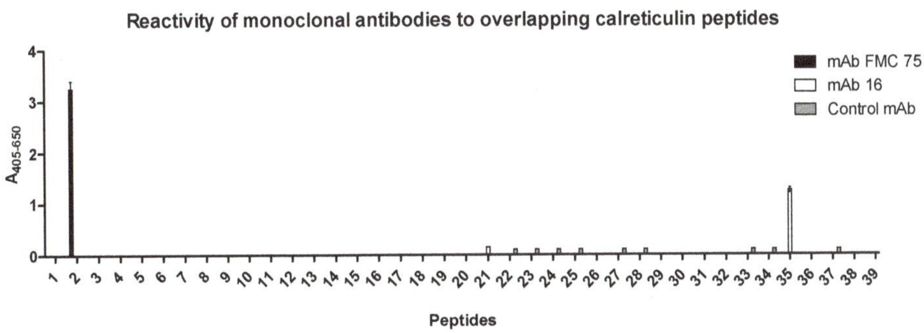


**Figure 2.** Antibody reactivity to overlapping calreticulin peptides analyzed by modified enzyme-linked immunosorbent assay using resin-bound peptides. Mouse anti-β-galactosidase was used as a negative control.

Based on the current findings, peptide 2 (LDGDGWTSRWIESKHKSDFG) and 35 (KDKQDEEQRLKEEEEDKKRK) were selected for epitope identification.

### 3.3. Identification of Minimal Sequences for Antibody Binding

To determine the minimal peptide length required for Ab binding, resin-bound peptides truncated from the N- and C-terminal ends were screened for Ab reactivity by modified ELISA. Peptides GDGWTSRWIESKHKSD (2) and KQDEEQRLKEEEEDKK (35) were used as templates for the generation of truncated peptides and functioned as the positive controls.

Figure 3a,b illustrates the reactivity of mAb FMC 75 to N- and C-terminally truncated peptides. As seen, the N-terminal aa Thr was crucial for Ab binding as no Ab reactivity was found to the SRWIESKHKSD peptide compared to the TSRWIESKHKSD peptide. Similarly, the C-terminal aa Lys was found to be crucial for Ab binding, as no reactivity was found to the GDGWTSRWIES peptide. Based on the current findings, the sequence TSRWIESK was determined to be the epitope of mAb FMC 75.

Figure 3c,d illustrate the reactivity of mAb 16 to N- and C-terminally truncated peptides. As seen, the terminal Asp in positions 3 and 14 of the 16-mer template peptide were crucial, respectively, as peptides that did not contain one of these aas did not react with the mAb 16. Based on the current findings, the sequence DEEQRLKEEED was identified as the epitope of mAb 16.

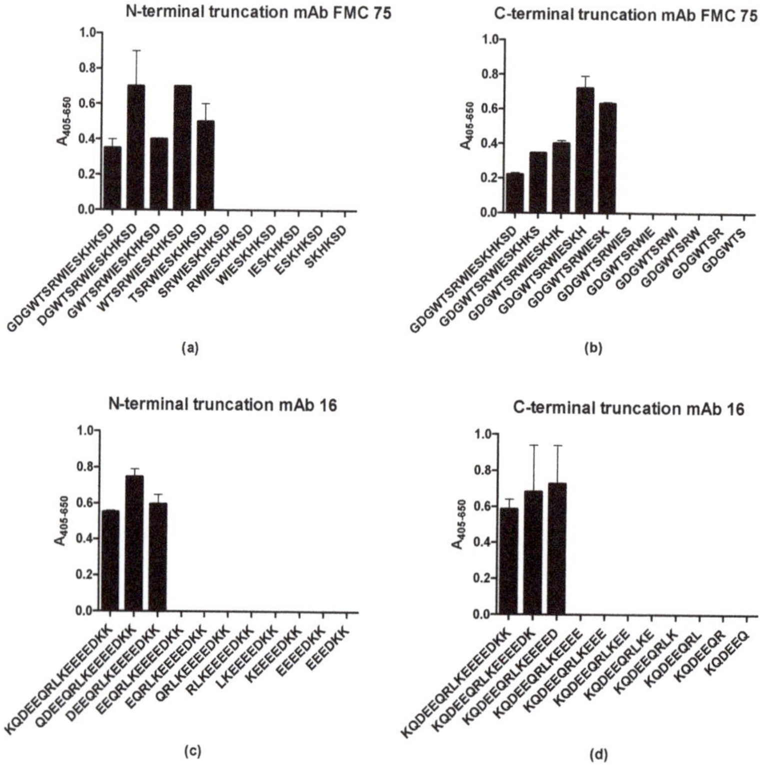


**Figure 3.** Antibody reactivity to N- and C-terminally truncated peptides. (**a**) Reactivity of mAb FMC 75 to N-terminal truncated peptides. (**b**) Reactivity of mAb FMC 75 to C-terminal truncated peptides. (**c**) Reactivity of mAb 16 to N-terminal truncated peptides. (**d**) Reactivity of mAb 16 to C-terminal truncated peptides.

### *3.4. Determination of Antigenic Amino Acids Essential for Antibody Reactivity*

The preliminary analysis identified the sequences TSRWIESK and DEEQRLKEEEED as the epitopes of mAb FMC 75 and mAb 16, respectively. To identify aas essential for binding in the identified epitopes, alanine scanning was conducted, where each aa in the identified epitopes of mAb FMC 75 and mAb 16 was substituted with alanine one at a time.

Figure 4a illustrates the reactivity of mAb FMC 75 to alanine-substituted peptides, using the peptide TSRWIESK as a template. As seen, the aas $Thr^1$, $Ser^2$, $Arg^3$, $Trp^4$, and $Lys^8$ identified relative to the template peptide were essential for Ab binding. The last three residues either have the same or higher intensity compared to the control peptide.

Figure 4b illustrates the reactivity of mAb 16 to alanine-substituted peptides analyzed by modified ELISA. As seen, the majority of the alanine-substituted peptides did not react with the mAb 16; only the substitution of $Glu^2$ and $Leu^6$ was tolerated. Furthermore, the Ab binding was reduced by approximately 40% when $Glu^{10}$ was substituted with Ala.

Next, to determine whether the dependency of the individual aa residue relates to the functional group in the side chain, a functionality scan was conducted, where the specific contributions of the crucial aa side chains were determined by replacing the aas with an aa of similar functionality.

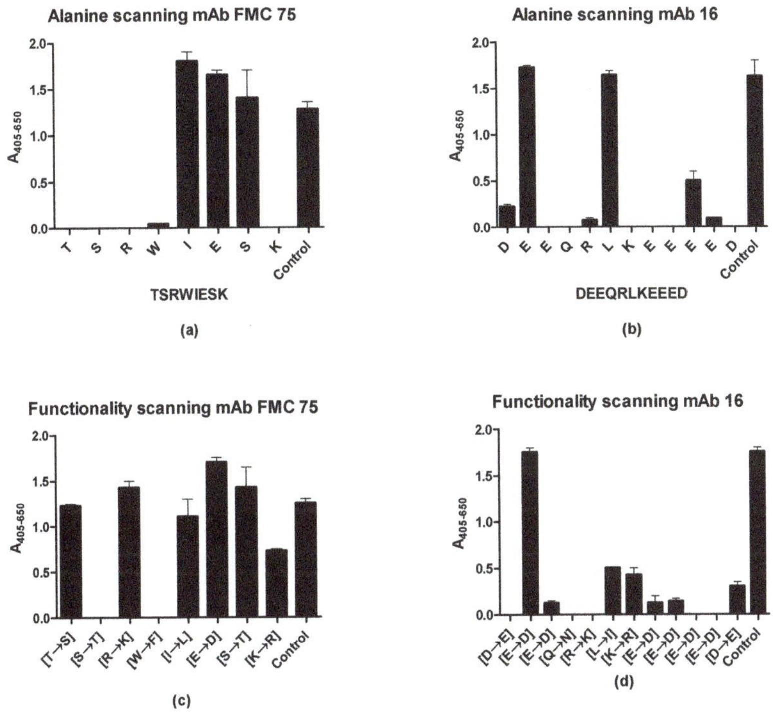


**Figure 4.** Antibody reactivity to alanine- and functionality-substituted peptides. (**a**) Reactivity of mAb FMC 75 to alanine-substituted peptides. (**b**) Reactivity of mAb 16 to alanine-substituted peptides. (**c**) Reactivity of mAb FMC 75 to functionality-substituted peptides. (**d**) Reactivity of mAb 16 to functionality-substituted peptides.

Figure 4c illustrates the reactivity of mAb FMC 75 to functionality-substituted peptides. As seen, Ab reactivity was reduced when $Ser^2$ was substituted with Thr and when $Trp^4$ was substituted with Phe. Furthermore, Ab reactivity to the $Lys^8$ was reduced by approximately 40% when substituted to Arg when compared with the positive control.

Figure 4d illustrates the reactivity of mAb 16 to functionality-substituted peptides. As seen, Ab reactivity was significantly reduced when replacing the majority of the aas compared to the control peptides. Only the substitution of $Glu^2$ with Asp did not reduce Ab binding, whereas the binding was reduced by 60–100% for the remaining residues relative to the control.

Collectively, the characterized Abs have a high dependency on the majority of the aas represented in the epitope.

### *3.5. Epitope Presentation in the Native Calreticulin Structure*

Finally, the native structures of the identified epitopes were determined using a CRT crystal structure covering aas 16–367, originally identified by Chouquet et al. [16].

As seen in Figure 5a, the native epitope of mAb FMC 75 (yellow) was found in a flexible β-strand structure. The essential aas Thr, Ser, Arg, and Lys in the mAb FMC 75 epitope protruded in the crystal structure, whereas the Trp residue pointed into the core structure of CRT (Figure 5b).

Unfortunately, the complete structure of CRT has not been crystallized, and hence only the N-terminal region of the mAb 16 epitope (aa 263–373) could be identified. However, the first five aas of the C-terminal epitope were identified to be located in an α-helix structure. Moreover, as presented in Figure 5c, α-helical wheel representation of mAb 16 indicates that the complete epitope is found in an α-helix structure, where residue 2, 6, and 10 were seen to be located on the same side of the potential α-helix.

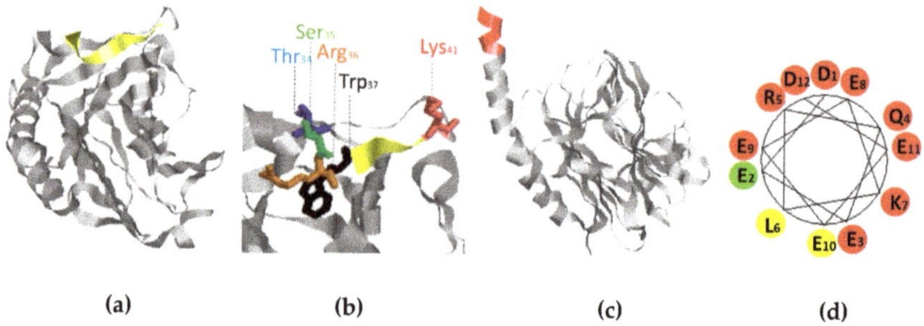


**Figure 5.** Visualization of the identified epitopes in the native structure of calreticulin. (**a**) Visualization of the mAb FMC 75 epitope. (**b**) Visualization of the essential amino acids in the mAb FMC 75 epitope (amino acids 34–). (**c**) Visualization of the N-terminal region (363–367) of the mAb 16 epitope (362–373). (**d**) α-helical wheel representation of mAb 16. "Red" represents aas which are critical for antibody binding. "Yellow" represents aas semi-critical aas. "Green" represents non-critical aas residues.

### *3.6. Reactivity of mAb FMC 75 to Oligomerized Calreticulin*

As presented, the epitope of FMC 75 was located in the N-terminal; in contrast, the epitope of mAb 16 was located in the C-terminal end, where potential fs mutations occasionally may occur. To determine whether the epitope of mAb FMC 75 has therapeutic potential, a Coomassie staining and a Western blot were conducted to examine whether the epitope of mAb FMC 75 is assessable upon oligomerization, as the oligomeric structure of CRT is essential for some of the protein's functions.

As presented in Figure 6a, CRT formed oligomers upon heat treatment, which were recognized by the mAb FMC 75 (Figure 6b), suggesting that the epitope of mAb FMC 75 could be a potential therapeutic target, as it is assessable in the oligomeric structure.

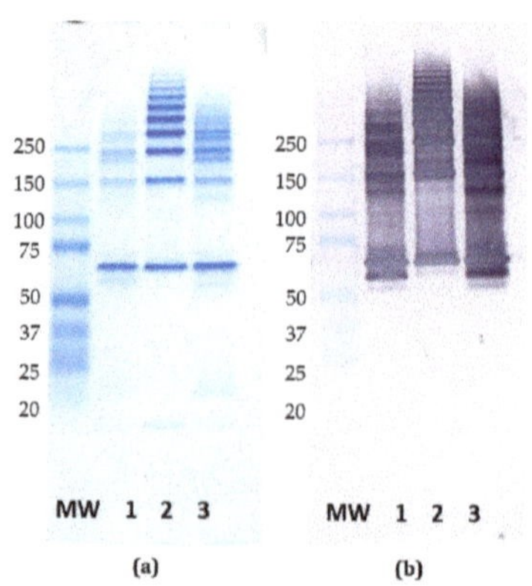


**Figure 6.** Native PAGE and Western blotting of wild-type calreticulin. (**a**) Coomassie staining of oligomerized calreticulin. (**b**) Western blotting of mAb FMC 75 to oligomerized calreticulin. MW: Molecular weight (KDa), lane 1: heat-treated human recombinant CRT (3 days at 42 °C, 57 °C 10 min), lane 2: heat-treated human CRT purified from human placenta (57 °C 10 min) lane 3; heat-treated human recombinant CRT (57 °C 10 min).

## 4. Discussion

This study describes the specific reactivity of mAb FMC 75 and mAb 16, which both have contributed to elucidate the biological function of CRT, among others, through detection of CRT exposure in the course of immunogenic cell death and examination of the biophysical interaction with other target molecules [33,34,38]. Moreover, FMC 75 has contributed to studies describing tumor-associated CRT variants [36].

In this study, both Abs recognized the wt CRT in ELISA, and as shown in Western blotting, the mAb FMC 75 recognized monomers as well as oligomers. These findings indicate that the epitope of the mAb FMC 75 is accessible upon oligomerization, which is critical for some CRT functions [19], and that the epitope of FMC 75 could be a potential therapeutic target, e.g., in myeloproliferative diseases. In contrast, the epitope of mAb 16 was identified to be located in the C-terminal end, which is prone to mutations causing fs in the C-terminus. Hence, the value of mAb 16 as a potential therapeutic is more limited.

The two peptide-specific CRT Abs recognized epitopes in the N- or C-terminal end of wt CRT, thus, the CRT mAbs reacted to the synthetic peptides as well as the native Ags (Figures 1 and 2), which is a crucial factor defining a well-functioning Ab [25–27]. Moreover, these findings indicate that the epitopes of the CRT mAbs are located in surface-accessible and flexible regions, which is typical for many Abs recognizing linear epitopes [40,42,43] and which is confirmed by comparison with published CRT structures [16,17].

Using overlapping peptides covering the complete CRT sequence, it was possible to identify the antigenic regions of mAb FMC 75 and mAb 16 (Figure 2), an approach which often has been applied with success for Ab characterization [42–45], as screening for Ab reactivity directly on the resin allows rapid identification [40–42].

Using N- and C-terminally truncated peptides, the essential terminal aas for both epitopes were determined, corresponding to eight residues in the N-terminus for mAb FMC 75 (34–41, TSRWIESK) and 12 residues in the C-terminus for mAb 16, corresponding to aa 362–373 (DEEQRLKEEEED) (Figure 3). For both epitopes, the charged aas were essential for binding. The two positively charged aas in the epitope of mAb FMC 75 (Arg and Lys) could not be substituted with Ala without reducing Ab binding; however, it was possible to substitute both with aas containing a positive charge and still obtain a significant level of reactivity, indicating that the presence of a positively charged aa rather than the specific aa was essential for Ab binding. Similar to mAb 16, charged aas appeared to prevail in the interaction, as most of the aas found in the epitope of mAb 16 were positively or negatively charged aas (Figure 4). Alanine substitution was only tolerated in positions 2 and 6 (Glu and Leu). As the peptide is believed to be in $\alpha$ -helical conformation, this fits with residues 2, 6, and 10 being located on the same side of the $\alpha$-helix as shown by an $\alpha$-helical wheel. However, when substituting Glu2 and Leu6 with aas of similar functionality (Gln and Ile, respectively), Ab reactivity was still obtained. Collectively, these findings indicated that Glu and Leu in positions 2 and 6, respectively, are not crucial for Ab binding and presumably are located in the part of the $\alpha$-helix contacting the rest of the molecule (i.e., away from the solvent). As presented in Figure 5c, the N-terminal region of the mAb 16 epitope was located in an $\alpha$-helix structure. Although the complete epitope could not be determined from the crystal structure, recent findings have proposed that the missing region of the epitope is found in an $\alpha$-helix structure as well [17,46,47]. Collectively, these findings contribute to increased knowledge of the crystal structure of the C-terminal end of CRT, as only the first 365 aas of CRT have been crystallized.

It remains to be determined whether all aas in the epitope are directly involved in the Ab- Ag interaction, as the aas side chains protrude in different directions around the $\alpha$-helix structure. The most reasonable explanation is that some of the charged aas make direct contact with the Ab, whereas the remaining aas function to stabilize the helix structure.

Whereas most of the aa residues of epitope 2 were essential for the Ab-Ag interaction, the aas of epitope 1 (34–41, TSRWIESK) were more tolerant to substitutions. Besides the dependency of the positively charged aas already mentioned, the epitope of mAb FMC 75 was found to depend on other factors for binding as well. Thus, Thr in position 1 appeared to contribute a hydroxy group for the formation of a hydrogen bond, as Thr could not be substituted with Ala but could be substituted with Ser. Similarly, Ser in position 2 could not be substituted with either Thr or Ala, indicating that the side-chain of Ser in position

2 is essential for Ab binding. Trp in position 4 could not be substituted with either Ala or Phe, indicating that this aa is essential for binding as well. However, when comparing the results to the crystal structure, it was found that the Trp residue points into the core structure of CRT, suggesting that Trp does not make direct contact with mAb FMC 75 but functions to stabilize the epitope. The aas in positions 5–7 could be substituted with Ala or aas with similar functionality, indicating that these aas contribute with backbone conformation and stability or hydrogen bonds to the Ab-epitope interface. Collectively, these findings indicate that charged aas and aas contributing with hydrogen bonds are essential for Ab binding.

The versatile nature of the specific contribution of the individual aas in the Ab-epitope interface is in accordance with the literature [40–45]. Several examples of characterized Ab-Ag interactions illustrate that the contribution of the epitope is unique for the specific complex [40–45]. Thus, some interactions appear to be backbone-dependent, some appear to depend on charged and hydrophilic aas, as described in this study, and some depend on hydrophobic aas for a stable interaction [40–45,48,49]. This clearly illustrates the versatile nature of Abs and Ab-Ag interactions, which is in accordance with the findings described in this study.

**Author Contributions:** Conceptualization, G.H., P.H., and N.H.T.; methodology, A.C.B., C.K., E.C., G.H., N.H.T., R.S.; software, N.H.T.; validation, A.C.B., C.K., G.H., and N.H.T.; formal analysis, G.H., A.C.B.; investigation, A.C.B., G.H.; resources, E.C., G.H., N.H.T., R.S.; data curation, A.C.B. and N.H.T.; writing—original draft preparation, N.H.T., and G.H.; writing—review and editing, N.H.T. and G.H.; N.H.T.; supervision, G.H.; project administration, G.H.; funding acquisition, G.H. All authors have read and agreed to the published version of the manuscript.

**Funding:** This research received no external funding.

**Institutional Review Board Statement:** Ethical review and approval were waived for this study, as commercial monoclonal antibodies and reagents primarily were used in the present study.

**Informed Consent Statement:** A human placenta used for calreticulin purification was obtained with informed consent.

**Data Availability Statement:** All the data obtained in the current study is presented in the manuscript.

**Conflicts of Interest:** The authors declare no conflict of interest.

## Appendix A

**Table A1.** Overlapping CRT Peptides Covering the CRT Protein.

| # | Amino Acid | Peptide |
|---|---|---|
| 1 | 18–37 | EPAVYFKEQFLDGDGWTSRW |
| 2 | 28–47 | LDGDGWTSRWIESKHKSDFG |
| 3 | 38–57 | IESKHKSDFGKFVLSSGKFY |
| 4 | 48–67 | KFVLSSGKFYGDEEKDKGLQ |
| 5 | 58–77 | GDEEKDKGLQTSQDARFYAL |
| 6 | 68–87 | TSQDARFYALSASFEPFSNK |
| 7 | 78–97 | SASFEPFSNKGQTLVVQFTV |
| 8 | 88–107 | GQTLVVQFTVKHEQNIDCGG |
| 9 | 98–117 | KHEQNIDCGGGYVKLFPNSL |
| 10 | 108–127 | GYVKLFPNSLDQTDMHGDSE |
| 11 | 118–137 | DQTDMHGDSEYNIMFGPDIC |
| 12 | 128–147 | YNIMFGPDICGPGTKKVHVI |
| 13 | 138–157 | GPGTKKVHVIFNYKGKNVLI |
| 14 | 148–167 | FNYKGKNVLINKDIRCKDDE |

**Table A1.** *Cont.*

| # | Amino Acid | Peptide |
|---|---|---|
| 15 | 158–177 | NKDIRCKDDEFTHLYTLIVR |
| 16 | 168–187 | FTHLYTLIVRPDNTYEVKID |
| 17 | 178–197 | PDNTYEVKIDNSQVESGSLE |
| 18 | 188–207 | NSQVESGSLEDDWDFLPPKK |
| 19 | 198–217 | DDWDFLPPKKIKDPDASKPE |
| 20 | 208–227 | IKDPDASKPEDWDERAKIDD |
| 21 | 218–237 | DWDERAKIDDPTDSKPEDWD |
| 22 | 228–247 | PTDSKPEDWDKPEHIPDPDA |
| 23 | 238–257 | KPEHIPDPDAKKPEDWDEEM |
| 24 | 248–267 | KKPEDWDEEMDGEWEPPVIQ |
| 25 | 258–277 | DGEWEPPVIQNPEYKGEWKP |
| 26 | 268–287 | NPEYKGEWKPRQIDNPDYKG |
| 27 | 278–297 | RQIDNPDYKGTWIHPEIDNP |
| 28 | 288–307 | TWIHPEIDNPEYSPDPSIYA |
| 29 | 298–317 | EYSPDPSIYAYDNFGVLGLD |
| 30 | 308–327 | YDNFGVLGLDLWQVKSGTIF |
| 31 | 318–337 | LWQVKSGTIFDNFLITNDEA |
| 32 | 328–347 | DNFLITNDEAYAEEFGNETW |
| 33 | 338–357 | YAEEFGNETWGVTKAAEKQM |
| 34 | 348–367 | GVTKAAEKQMKDKQDEEQRL |
| 35 | 358–377 | KDKQDEEQRLKEEEEDKKRK |
| 36 | 368–387 | KEEEEDKKRKEEEEAEDKED |
| 37 | 378–397 | EEEEAEDKEDDEDKDEDEED |
| 38 | 388–407 | DEDKDEDEEDEEDKEEDEEE |
| 39 | 398–417 | EEDKEEDEEEDVPGQAKDEL |

## References

1. Wang, W.-A.; Groenendyk, J.; Michalak, M. Calreticulin signaling in health and disease. *Int. J. Biochem. Cell Biol.* **2012**, *44*, 842–846. [CrossRef]
2. Michalak, M.; Groenendyk, J.; Szabo, E.; Gold, L.I.; Opas, M. Calreticulin, a multi-process calcium-buffering chaperone of the endoplasmic reticulum. *Biochem. J.* **2009**, *417*, 651–666. [CrossRef]
3. Martins, I.; Kepp, O.; Galluzzi, L.; Senovilla, L.; Schlemmer, F.; Adjemian, S.; Menger, L.; Michaud, M.; Zitvogel, L.; Kroemer, G. Surface-exposed calreticulin in the interaction between dying cells and phagocytes. *Ann. N. Y. Acad. Sci.* **2010**, *1209*, 77–82. [CrossRef] [PubMed]
4. Holoshitz, J.; De Almeida, D.E.; Ling, S. A role for calreticulin in the pathogenesis of rheumatoid arthritis. *Ann. N. Y. Acad. Sci.* **2010**, *1209*, 91–98. [CrossRef] [PubMed]
5. Gold, L.I.; Eggleton, P.; Sweetwyne, M.T.; Van Duyn, L.B.; Greives, M.R.; Naylor, S.; Michalak, M.; Murphy-Ullrich, J.E. Calreticulin: Non-endoplasmic reticulum functions in physiology and disease. *FASEB J.* **2010**, *24*, 665–683. [CrossRef] [PubMed]
6. Kumari, R.; Bansal, S.; Gupta, G.; Arora, S.; Kumar, A.; Goel, S.; Singh, P.; Ponnan, P.; Priya, N.; Tyagi, T.K.; et al. Calreticulin transacylase: Genesis, mechanism of action and biological applications. *Biochimie* **2010**, *92*, 1173–1179. [CrossRef]
7. Villagomez, M.; Szabo, E.; Podcheko, A.; Feng, T.; Papp, S.; Opas, M. Calreticulin and focal-contact-dependent adhesion. *Biochem. Cell Biol.* **2009**, *87*, 545–556. [CrossRef]
8. Chaput, N.; De Botton, S.; Obeid, M.; Apetoh, L.; Ghiringhelli, F.; Panaretakis, T.; Flament, C.; Zitvogel, L.; Kroemer, G. Molecular determinants of immunogenic cell death: Surface exposure of calreticulin makes the difference. *J. Mol. Med.* **2007**, *85*, 1069–1076. [CrossRef]
9. Imai, M.; Araki, M.; Komatsu, N. Somatic mutations of calreticulin in myeloproliferative neoplasms. *Int. J. Hematol.* **2017**, *105*, 743–747. [CrossRef]
10. Clinton, A.; McMullin, M.F. The Calreticulin gene and myeloproliferative neoplasms. *J. Clin. Pathol.* **2016**, *69*, 841–845. [CrossRef]
11. Chi, J.; Manoloukos, M.; Pierides, C.; Nicolaidou, V.; Nicolaou, K.; Kleopa, M.; Vassiliou, G.; Costeas, P. Calreticulin mutations in myeloproliferative neoplasms and new methodology for their detection and monitoring. *Ann. Hematol.* **2015**, *94*, 399–408. [CrossRef]
12. Klampfl, T.; Gisslinger, H.; Harutyunyan, A.; Nivarthi, H.; Rumi, E.; Milosevic, J.D.; Them, N.C.; Berg, T.; Gisslinger, B.; Pietra, D.; et al. Somatic Mutations of Calreticulin in Myeloproliferative Neoplasms. *N. Engl. J. Med.* **2013**, *369*, 2379–2390. [CrossRef]
13. Nangalia, J.; Massie, C.E.; Baxter, E.J.; Nice, F.L.; Gundem, G.; Wedge, D.C.; Avezov, E.; Li, J.; Kollmann, K.; Kent, D.G.; et al. Somatic CALR mutations in myeloproliferative neoplasms with nonmutated JAK2. *N. Engl. J. Med.* **2013**, *369*, 2391–2405. [CrossRef] [PubMed]

14. Kozlov, G.; Gehring, K. Calnexin cycle—Structural features of the ER chaperone system. *FEBS J.* **2020**, *287*, 4322–4340. [CrossRef] [PubMed]
15. Kozlov, G.; Pocanschi, C.L.; Rosenauer, A.; Bastos-Aristizabal, S.; Gorelik, A.; Williams, D.B.; Gehring, K. Structural Basis of Carbohydrate Recognition by Calreticulin. *J. Biol. Chem.* **2010**, *285*, 38612–38620. [CrossRef] [PubMed]
16. Chouquet, A.; Païdassi, H.; Ling, W.L.; Frachet, P.; Houen, G.; Arlaud, G.J.; Gaboriaud, C. X-ray structure of the human calreticulin globular domain reveals a peptide-binding area and suggests a multi-molecular mechanism. *PLoS ONE* **2011**, *6*, e17886. [CrossRef] [PubMed]
17. Boelt, S.G.; Norn, C.; Rasmussen, M.I.; André, I.; Čiplys, E.; Slibinskas, R.; Houen, G.; Højrup, P. Mapping the Ca(2+) induced structural change in calreticulin. *J. Proteomics* **2016**, *142*, 138–148. [CrossRef] [PubMed]
18. Houen, G.; Højrup, P.; Ciplys, E.; Gaboriaud, C.; Slibinskas, R. Structural Analysis of calreticulin, an endoplasmic reticulum-resident molecular chaperone. In *Cellular Biology of the Endoplasmic Reticulum*; Springer: Basel, Switzerland, 2021; pp. 13–25.
19. Jørgensen, C.S.; Ryder, L.R.; Steinø, A.; Højrup, P.; Hansen, J.; Beyer, N.H.; Heegaard, N.H.H.; Houen, G. Dimerization and oligomerization of the chaperone calreticulin. *Eur. J. Biochem.* **2003**, *270*, 4140–4148. [CrossRef]
20. Rizvi, S.M.; Mancino, L.; Thammavongsa, V.; Cantley, R.L.; Raghavan, M. A Polypeptide Binding Conformation of Calreticulin Is Induced by Heat Shock, Calcium Depletion, or by Deletion of the C-Terminal Acidic Region. *Mol. Cell* **2004**, *15*, 913–923. [CrossRef]
21. Huang, S.-H.; Zhao, L.-X.; Hong, C.; Duo, C.-C.; Guo, B.-N.; Zhang, L.-J.; Gong, Z.; Xiong, S.-D.; Gong, F.-Y.; Gao, X.-M. Self-Oligomerization Is Essential for Enhanced Immunological Activities of Soluble Recombinant Calreticulin. *PLoS ONE* **2013**, *8*, e64951. [CrossRef]
22. Araki, M.; Yang, Y.; Imai, M.; Mizukami, Y.; Kihara, Y.; Sunami, Y.; Masubuchi, N.; Edahiro, Y.; Hironaka, Y.; Osaga, S.; et al. Homomultimerization of mutant calreticulin is a prerequisite for MPL binding and activation. *Leukemia* **2019**, *33*, 122–131. [CrossRef] [PubMed]
23. Rivera, J.F.; Baral, A.J.; Nadat, F.; Boyd, G.; Smyth, R.; Patel, H.; Burman, E.L.; Alameer, G.; Boxall, S.A.; Jackson, B.R.; et al. Zinc-dependent multimerization of mutant calreticulin is required for MPL binding and MPN pathogenesis. *Blood Adv.* **2021**, *5*, 1922–1932. [CrossRef] [PubMed]
24. Venkatesan, A.; Geng, J.; Kandarpa, M.; Wijeyesakere, S.J.; Bhide, A.; Talpaz, M.; Pogozheva, I.D.; Raghavan, M. Mechanism of mutant calreticulin-mediated activation of the thrombopoietin receptor in cancers. *J. Cell Biol.* **2021**, *220*, e202009179. [CrossRef]
25. Houen, G.; Trier, N. Molecular Recognition and Advances in Antibody Design and Antigenic Peptide Targeting. *Int. J. Mol. Sci.* **2020**, *21*, 1405. [CrossRef] [PubMed]
26. Trier, N.H.; Houen, G. Antibodies as Diagnostic Targets and as Reagents for Diagnostics. *Antibodies* **2020**, *9*, 15. [CrossRef] [PubMed]
27. Trier, N.; Hansen, P.; Houen, G. Peptides, antibodies, peptide antibodies and more. *Int. J. Mol. Sci.* **2019**, *20*, 6289. [CrossRef]
28. Trier, N.H.; Houen, G. Peptide antibodies in clinical laboratory diagnostics. *Adv. Clin. Chem.* **2017**, *81*, 43–96.
29. Wang, G.; Yang, J.; Liu, C. A mouse monoclonal antibody specific for calreticulin. *Hybridoma* **2012**, *31*, 382–385. [CrossRef]
30. Corsiero, E.; Jagemann, L.; Perretti, M.; Pitzalis, C.; Bombardieri, M. Characterization of a synovial B cell-derived recombinant monoclonal antibody targeting stromal calreticulin in the rheumatoid joints. *J. Immunol.* **2018**, *201*, 1373–1381. [CrossRef]
31. Stein, H.; Bob, R.; Dürkop, H.; Erck, C.; Kämpfe, D.; Kvasnicka, H.M.; Martens, H.; Roth, A.; Streubel, A. A new monoclonal antibody (CAL2) detects calreticulin mutations in formalin-fixed and paraffin-embedded bone marrow biopsies. *Leukemia* **2016**, *30*, 131–135. [CrossRef] [PubMed]
32. Pandya, U.M.; Manzanares, M.A.; Tellechea, A.; Egbuta, C.; Daubriac, J.; Jimenez-Jaramillo, C.; Samra, F.; Fredston-Hermann, A.; Saadipour, K. Calreticulin exploits TGF-β for extracellular matrix induction engineering a tissue regenerative process. *FASEB J.* **2020**, *34*, 15849–15874. [CrossRef]
33. Kasikova, L.; Truxova, I.; Cremer, I.; Sautes-Fridman, C.; Kepp, O.; Kroemer, G.; Spisek, R.; Fucikova, J. Side-by-side comparison of flow cytometry and immunohistochemistry for detection of calreticulin exposure in the course of immunogenic cell death. *Methods Enzym.* **2020**, *632*, 15–25.
34. Pandya, U.M.; Egbuta, C.; Norman, T.F.A.; Chiang, C.; Wiersma, V.R.; Panchal, R.G.; Bremer, E.; Eggleton, P.; Gold, L.I. The biophysical interaction of the danger-associated molecular pattern (DAMP) calreticulin with the pattern-associated molecular pattern (PAMP) lipopolysaccharide. *Int. J. Mol. Sci.* **2019**, *20*, 408. [CrossRef] [PubMed]
35. Tu, L.; Wang, J.; Barathi, V.A.; Prea, S.M.; He, Z.; Lee, J.H.; Bender, J.; King, A.E.; Logan, G.J.; Alexander, I.E.; et al. AAV-mediated gene delivery of the calreticulin anti-angiogenic domain inhibits ocular neovascularization. *Angiogenesis* **2018**, *21*, 95–109. [CrossRef] [PubMed]
36. Arshad, N.; Cresswell, P. Tumor-associated calreticulin variants functionally compromise the peptide loading complex and impair its recruitment of MHC-I. *J. Biol. Chem.* **2018**, *293*, 9555–9569. [CrossRef]
37. Ito, H.; Seyama, Y.; Kubota, S. Calreticulin is directly involved in anti-alpha3 integrin antibody-mediated secretion and activation of matrix metalloprotease-2. *Biochem. Biophys. Res. Commun.* **2001**, *283*, 297–302. [CrossRef]
38. Nair, S.; Wearsch, P.A.; Mitchell, D.A.; Wassenberg, J.J.; Gilboa, E.; Nicchitta, C.V. Calreticulin displays in vivo peptide-binding activity and can elicit CTL responses against bound peptides. *J. Immunol.* **1999**, *162*, 6426–6432.
39. Čiplys, E.; Žitkus, E.; Gold, L.I.; Daubriac, J.; Pavlides, S.C.; Højrup, P.; Houen, G.; Wang, W.A.; Michalak, M.; Slibinskas, R. High-level secretion of native recombinant human calreticulin in yeast. *Microb. Cell Fact.* **2015**, *14*, 165. [CrossRef]

40. Amrutkar, S.D.; Trier, N.H.; Hansen, P.R.; Houen, G. Fine mapping of a monoclonal antibody to the N-Methyl D-aspartate receptor reveals a short linear epitope. *Biopolymers* **2012**, *98*, 567–575. [CrossRef]
41. Petersen, N.H.; Hansen, P.R.; Houen, G. Fast and efficient characterization of an anti-gliadin monoclonal antibody epitope related to celiac disease using resin-bound peptides. *J. Immunol. Methods* **2011**, *365*, 174–182. [CrossRef]
42. Valdarnini, N.; Holm, B.; Hansen, P.; Rovero, P.; Houen, G.; Trier, N. Fine mapping of glutamate decarboxylase 65 epitopes reveals dependency on hydrophobic amino acids for specific interactions. *Int. J. Mol. Sci.* **2019**, *20*, 2909. [CrossRef]
43. Trier, N.H.; Hansen, P.R.; Vedeler, C.A.; Somnier, F.E.; Houen, G. Identification of continuous epitopes of HuD antibodies related to paraneoplastic diseases/small cell lung cancer. *J. Neuroimmunol.* **2012**, *243*, 25–33. [CrossRef] [PubMed]
44. Welner, S.; Trier, N.H.; Houen, G.; Hansen, P.R. Identification and mapping of a linear epitope of centromere protein F using monoclonal antibodies. *J. Pept. Sci.* **2013**, *19*, 95–101. [CrossRef]
45. Agca, S.; Houen, G.; Trier, N.H. Characterization of continuous B-cell epitopes in the N-terminus of glutamate decarboxylase67 using monoclonal antibodies. *J. Pept. Sci.* **2014**, *20*, 928–934. [CrossRef] [PubMed]
46. Fisette, O.; Schröder, G.F.; Schäfer, L.V. Atomistic structure and dynamics of the human MHC-I peptide loading complex. *Proc. Natl. Acad. Sci. USA* **2020**, *117*, 20597–20606. [CrossRef] [PubMed]
47. Trier, N.H.; Holm, B.E.; Slot, O.; Locht, H.; Lindegaard, H.; Svendsen, A.; Houen, G. Physical characteristics of a citrullinated pro-filaggrin epitope recognized by anti-citrullinated protein antibodies in rheumatoid arthritis sera. *PLoS ONE* **2016**, *11*, e0168542. [CrossRef] [PubMed]
48. Blees, A.; Januliene, D.; Hofmann, T.; Koller, N.; Schmidt, C.; Trowitzsch, S.; Moeller, A.; Tampé, R. Structure of the human MHC-I peptide-loading complex. *Nature* **2017**, *551*, 525–528. [CrossRef]
49. Trier, N.H.; Dam, C.E.; Olsen, D.T.; Hansen, P.R.; Houen, G. Contribution of peptide backbone to citrulline-dependent antibody reactivity. *PLoS ONE* **2015**, *10*, e0144707. [CrossRef]

antibodies

MDPI

*Article*

# Specificity of Anti-Citrullinated Protein Antibodies to Citrullinated α-Enolase Peptides as a Function of Epitope Structure and Composition

**Ilaria Fanelli [1], Paolo Rovero [1], Paul Robert Hansen [2], Jette Frederiksen [3], Gunnar Houen [3,4] and Nicole Hartwig Trier [3,*]**

1 Interdepartmental Laboratory of Peptide and Protein Chemistry and Biology, Department of NeuroFarBa, University of Florence, 50019 Sesto Fiorentino, Italy; ilaria.fanelli1@stud.unifi.it (I.F.); paolo.rovero@unifi.it (P.R.)
2 Department of Drug Design and Pharmacology, University of Copenhagen, 2100 Copenhagen, Denmark; prh@sund.ku.dk
3 Department of Neurology, Rigshospitalet Glostrup, 2600 Glostrup, Denmark; jette.lautrup.battistini@regionh.dk (J.F.); gunnar.houen@regionh.dk (G.H.)
4 Department of Biochemistry and Molecular Biology, University of Southern Denmark, 5230 Odense, Denmark
* Correspondence: nicole.hartwig.trier@regionh.dk

**Abstract:** Rheumatoid arthritis (RA) is an autoimmune disease affecting approximately 1–2% of the world population. In addition to the first discovered serologic markers for RA, the rheumatoid factors (RFs), anti-citrullinated protein antibodies (ACPAs) are even more specific for the disease compared to RFs and are found in 70–80% of RA patient sera. RA etiopathogenesis still needs to be elucidated, as different factors are proposed to be involved, such as Epstein–Barr virus infection. Hence, understanding the interaction between ACPAs and their citrullinated peptide targets is relevant for a better knowledge of RA pathophysiology and for diagnostic purposes. In this study, a cohort of RA sera, healthy control sera and multiple sclerosis sera were screened for reactivity to a variety of citrullinated peptides originating from α-enolase, pro-filaggrin, proteoglycan and Epstein–Barr nuclear antigen-2 by enzyme-linked immunosorbent assay. ACPA reactivity to citrullinated α-enolase peptides was found to depend on peptide length and peptide conformation, favouring cyclic (disulfide bond) conformations for long peptides and linear peptides for truncated ones. Additional investigations about the optimal peptide conformation for ACPA detection, employing pro-filaggrin and EBNA-2 peptides, confirmed these findings, indicating a positive effect of cyclization of longer peptides of approximately 20 amino acids. Moreover, screening of the citrullinated peptides confirmed that ACPAs can be divided into two groups based on their reactivity. Approximately 90% of RA sera recognize several peptide targets, being defined as cross-reactive or overlapping reactivities, and whose reactivity to the citrullinated peptide is considered primarily to be backbone-dependent. In contrast, approximately 10% recognize a single target and are defined as nonoverlapping, primarily depending on the specific amino acid side-chains in the epitope for a stable interaction. Collectively, this study contributed to characterize epitope composition and structure for optimal ACPA reactivity and to obtain further knowledge about the cross-reactive nature of ACPAs.

**Keywords:** anti-citrullinated protein antibodies; citrullinated peptides; epitopes; rheumatoid arthritis

**Citation:** Fanelli, I.; Rovero, P.; Hansen, P.R.; Frederiksen, J.; Houen, G.; Trier, N.H. Specificity of Anti-Citrullinated Protein Antibodies to Citrullinated α-Enolase Peptides as a Function of Epitope Structure and Composition. *Antibodies* **2021**, *10*, 27. https://doi.org/10.3390/antib10030027

Academic Editor: Ohad Mazor

Received: 22 May 2021
Accepted: 14 July 2021
Published: 21 July 2021



## 1. Introduction

RA is a systemic and chronic autoimmune disease with a worldwide prevalence of approximately 5 per 1000 adults, affecting women two to three times more often than men. RA disease onset may occur at any age; however, the peak incidence is in the sixth decade [1–3]. RA is characterized by infiltration of monocytes, B cells and T cells in the synovial membrane in joints and ultimately cartilage degradation and erosion of the underlying bone [2]. In addition to joint damage, some systemic features are associated with RA, for instance pulmonary, cardiovascular, psychological, and skeletal disorders [4,5]. Hence, RA has increased morbidity and mortality rates when left untreated [1,6].

RA is diagnosed according to EULAR/ACR classification criteria revised in 2010 which, along with clinical disease manifestations, comprise serological biomarkers such as anti-citrullinated protein antibodies (ACPA) [7]. ACPAs are detected in 70–80% of RA patients and in approximately 1–2% of the healthy population. Moreover, they have been reported to be present up to 14 years before the manifestation of clinical symptoms, making ACPAs good biomarkers for RA [8,9]. Ultimately, it has been reported that ACPA-positive RA patients experience increased joint damage and low remission rates, indicating that these individuals have more severe disease courses compared to ACPA-negative RA patients [10]. The occurrence of ACPA-positive RA is related to genetic risk factors that predispose for RA, for instance, protein tyrosine phosphatase nonreceptor type-22 (PTPN22) and MHC class II alleles [11–14].

ACPAs recognize the nonstandard amino acid citrulline, a nongenetically encoded amino acid. Citrullination is the result of a post-translational modification, where the positively charged guanidino group of Arg is substituted by the neutral ureido group. Ultimately, this modification may lead to structural unfolding of the citrullinated protein [15,16]. Citrullination is catalyzed by Peptidyl Arginine Deiminase (PAD) enzymes, which are calcium-dependent metalloenzymes [17].

ACPAs are typically detected with assays, which exploit enzyme-linked immunosorbent assay (ELISA) methods and synthetic citrullinated peptides [18–22]. The first generation of assays was based on a synthetic linear citrullinated peptide derived from human filaggrin [19]. In order to improve assay sensitivity, the linear peptide was replaced by a cyclic version (Cyclic Citrullinated Peptide, CCP, containing a disulfide bond), as the cyclic peptide yielded higher sensitivity and specificity compared to the linear version. This assay is also referred to as CCP1 [20]. Screening of peptide libraries has led to the selection of other antigens and generation of second and third generations of CCP assays [18]. While the previously mentioned assays only detect IgG ACPAs, the CCP3.1 detects both IgG and IgA isotypes. Despite this, the golden standard for ACPAs detection is the CCP2 assay [18,22].

In the commercial ACPA assays, different citrullinated peptides are employed, which is in accordance with the cross-reactive nature of ACPAs. ACPAs are able to recognize several citrullinated targets, preferably containing a Cit–Gly motif [19,21,23–27]. Besides a critical Cit–Gly motif, charged amino acids in the C-terminal have been proposed to be essential for a stable interaction between ACPAs and citrullinated peptide targets [23]. The amino acids surrounding citrulline have been analyzed in several studies, which revealed that substitutions in positions -x-x-Cit-Gly-x- do not influence antibody reactivity. This finding demonstrates the crucial role of the Cit–Gly motif for a stable antibody–antigen interaction, even though sometimes other amino acids besides Gly are also tolerated [23,24,26–28]. Examples of ACPA targets that have been reported are collagen, fibrinogen, α-enolase, vimentin, pro-filaggrin, Epstein–Barr nuclear antigen (EBNA)-1, and EBNA-2 [21,22,25].

It has been proposed that ACPAs can be divided into two groups, based on their ability to interact with citrullinated peptides [16,17], one group that appears to recognize a large variety of citrullinated targets and another group that recognizes a very limited number of citrullinated peptides [29]. The first group of ACPAs, also referred to as "overlapping" or "cross-reactive" antibodies, is primarily backbone-dependent, whereas the second group, also referred to as "nonoverlapping" or "epitope-specific", depends on the specific amino acid side-chains of the epitope to establish a stable antibody–antigen interaction [16].

On this basis, we analyzed the interactions between citrullinated targets and ACPAs in order to obtain further knowledge about ACPAs, which is important for the improvement of diagnostic tools, and to elucidate their role in the pathogenesis of RA. Citrullinated α-enolase peptides were used as a point of origin to characterize epitope composition and structure for optimal antibody reactivity and the overlapping and nonoverlapping ACPA reactivities.

## 2. Materials and Methods

### 2.1. Reagents

Alkaline phosphatase (AP)-conjugated goat-anti-human IgG, streptavidin and AP substrate tablets (*para*-nitrophenylphosphate (*p*NPP)) were from Sigma Aldrich (St. Louis, MO, USA). Tris-Tween-NaCl (TTN, 0.3 M NaCl, 20 mM Tris, 0.01% Tween 20, pH 7.5), carbonate buffer (0.05 M sodium carbonate, pH 9.6) and AP substrate buffer (1 M ethanolamine, 0.5 mM $MgCl_2$, pH 9.8), were from Statens Serum Institut (Copenhagen, Denmark). Synthetic peptides purchased were from Schäfer-N (Lyngby, Denmark) (Table 1) and were generated on TentaGel resin using standard Fmoc-based solid-phase peptide synthesis. The peptides were synthesized as peptide acids.

**Table 1.** Synthetic peptides tested for antibody reactivity. "B" represents biotin.

| Origin | Name | Sequence |
|---|---|---|
| α-enolase | C-10-Cit | CFDS-Cit-GNPTC |
| | L-10-Cit | CFDS-Cit-GNPTC |
| | C-12-Cit | CIFDS-Cit-GNPTVC |
| | L-12-Cit | CIFDS-Cit-GNPTVC |
| | C-14-Cit | CEIFDS-Cit-GNPTVEC |
| | L-14-Cit | CEIFDS-Cit-GNPTVEC |
| | C-19-Cit | KIHARCEIFDS-Cit-GNPTVEC |
| | L-19-Cit | KIHARCEIFDS-Cit-GNPTVEC |
| | L-19-Arg | KIHARCEIFDSRGNPTVEC |
| Fibronectin | Fibronectin L | DHEGTHSTK-Cit-GHAKSRPVRD(K(B)) |
| Proteoglycan | Proteoglycan L | B-PQASVPLRLT-Cit-GSRAPISRAQ |
| Pro-filaggrin | Pro-filaggrin C | HQCHQEST-Cit-GRSRGRCGRSGS(K(B)) |
| | Pro-filaggrin L | HQSHQEST-Cit-GRSRGRSGRSGS(K(B)) |
| Epstein–Barr virus | EBNA-2-L | GQGRGRWRG-Cit-GRSKGRGRMH(K(B)) |
| Nuclear antigen 2 | EBNA-2-C | GQGRCGRWRG-Cit-GRSKGRGCRMH(K(B)) |

### 2.2. Patient Material

RA serum samples ($n$ = 28) and healthy donor ($n$ = 28) serum samples (referred to as healthy control (HC)) were provided by Statens Serum Institut Biobank (Copenhagen, Denmark) ($n$ = 28), which routinely analyzes patient sera for diagnostic purposes. The samples were tested anonymously, therefore not requiring ethical consent.

Ten multiple sclerosis serum samples from the Multiple Sclerosis Clinic, Department of Neurology, Rigshospitalet Glostrup (Glostrup, Denmark) were used as disease controls. The samples were tested anonymously, therefore not requiring ethical consent.

### 2.3. Detection of Antibodies by Enzyme-Linked Immunosorbent Assay and Streptavidin-Capture Enzyme-Linked Immunosorbent Assay

Microtiter plates were coated with 1 µg/mL free peptide in carbonate buffer and incubated overnight at room temperature (RT) on a shaking table (ST). The wells were rinsed with TTN for 3 × 1 min and blocked with TTN for 30 min. Sera were diluted (1:200) in TTN, added to each well, and then incubated for 1 h (h) at RT on a ST. After washing with TTN buffer, AP-conjugated goat-anti-human IgG diluted in TTN (1:1000) was added to each well and incubated for 1h at RT on a ST. Finally, *p*NPP-containing AP substrate buffer (1 mg/mL) was added to each well and AP activity was determined by measuring the absorbance at 405 nm with background subtraction at 650 nm.

Alternatively, microtiter plates were precoated with 1 µg/mL streptavidin in carbonate buffer and incubated overnight at 4 °C. Biotinylated peptides (diluted to 1 µg/mL in carbonate buffer) were added to each well and incubated for 2h at RT on a ST. The following steps in the experiment were carried out as mentioned above. All samples were tested in duplicates.

Based on preliminary screening, absorbances of all the results were normalized to a positive RA control pool ($n = 28$) and a peptide-specific cutoff was introduced, tolerating a nonspecific reactivity of 5% and an intra-assay variation of 15%. Readings above the cutoff were regarded as positive, whereas samples below the cutoff were regarded as being negative. Inter-assay variations below 15% were acceptable.

### 2.4. Statistics

Statistical analyses and plots were generated using GraphPad Prism 9.0 software. The values obtained in the experiments were compared further by using Student's *t*-test.

## 3. Results

### 3.1. Reactivity of Rheumatoid Arthritis Sera to α-Enolase Peptides

Various citrullinated protein targets have been identified in RA, such as α-enolase, pro-filaggrin, proteoglycan, and fibronectin [30,31]. In order to further characterize the reactivity of ACPA to citrullinated peptides, RA patient sera ($n = 28$), HC sera ($n = 28$) and MS sera ($n = 10$) were tested for reactivity to a citrullinated α-enolase peptide (KIHARCEIFDS-Cit-GNPTVEC) by ELISA.

As seen in Figure 1, elevated antibody reactivity was found to the citrullinated peptides compared to the Arg-containing control peptide ($p = 0.0073$ for the cyclic and $p = 0.0003$ for the linear). No significant difference in antibody reactivity was found between the cyclic and the linear α-enolase peptides ($p = 0.2647$). Approximately 40% of the RA sera recognized the α-enolase peptides, and reacted significantly to the linear and the cyclic peptide compared to the control peptide ($p < 0.0001$). None of HC sera or MS sera reacted to the citrullinated α-enolase peptides, confirming that ACPA reactivity to the α-enolase peptides was specific for RA.

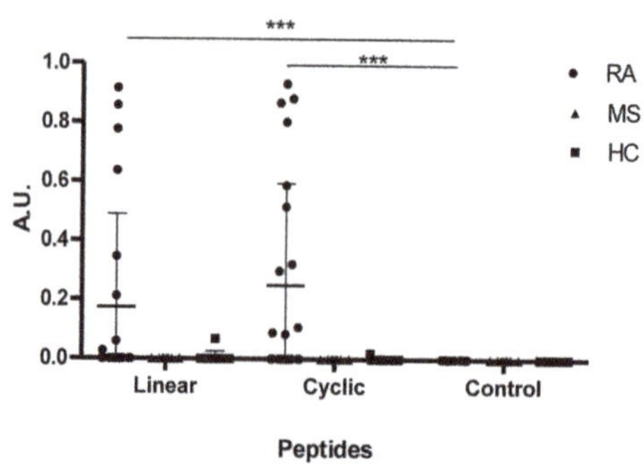


**Figure 1.** Reactivity of rheumatoid arthritis (RA), healthy control (HC) and multiple sclerosis (MS) sera to α-enolase peptides (KIHARCEIFDS-Cit-GNPTVEC) analysed by traditional ELISA. A linear Arg-containing peptide (KIHARCEIFDS-R-GNPTVEC) was used as negative control. HC and MS sera were used as controls. A.U. were defined as absorbances normalized relative to a positive RA control pool. *p* values less than 0.001 are shown as ***.

### 3.2. Reactivity of Rheumatoid Arthritis Sera to Truncated Linear and Cyclic α-Enolase Peptides

Previous studies describing RA sera reactivity to citrullinated pro-filaggrin peptides indicated that ACPA reactivity is dependent on peptide length and conformation [23]. To determine whether this relates to α-enolase peptides as well, RA sera that were positive for reactivity to α-enolase in the preliminary screening ($n = 12$) were tested for reactivity to cyclic and linear truncated α-enolase peptides by ELISA.

As seen in Figure 2, RA sera recognized all of the linear peptides independent of their length. Sensitivities of approximately 90% were found for all of the linear peptides. In contrast, specific ACPA reactivity was primarily found to the cyclic peptides C-19-Cit and C-10-Cit, obtaining sensitivities of approximately 90% as well. These findings indicate that the peptide conformation affects the antibody reactivity and that it is essential for peptide presentation. Additionally, no specific reactivity was found when screening HC sera. These findings are in accordance with the literature, describing that ACPA reactivity to 19-mer linear and cyclic pro-filaggrin peptides yields similar reactivity, whereas ACPA reactivity to the linear citrullinated peptides is favoured for smaller peptides (<12 amino acids), when compared to the cyclic peptides [23].

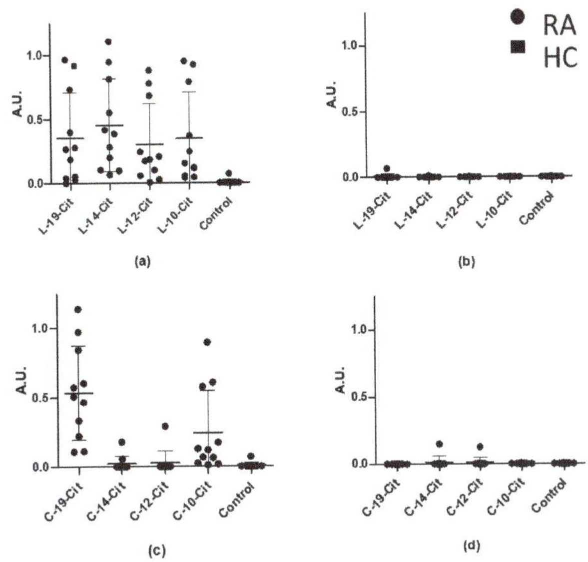


**Figure 2.** Reactivity of rheumatoid arthritis (RA) sera and healthy control (HC) sera samples to truncated linear and cyclic citrullinated α-enolase peptides analysed in ELISA. Peptides are plotted from left to right with decreasing length. An Arg-containing peptide was used as control (KIHARCEIFDSRGNPTVEC). (**a**) Reactivity of RA sera to linear peptides. (**b**) Reactivity of HC sera to linear peptides. (**c**) Reactivity of RA sera to cyclic peptides. (**d**) Reactivity of HC sera to cyclic peptides. A.U. were defined as absorbances normalized relative to a positive RA control pool.

### *3.3. Overlapping Reactivities of Anti-Citrullinated Protein Antibody Responses*

As presented, approximately 40% of the RA sera reacted with the α-enolase peptides. To determine whether the ACPA reactivities were specific for the α-enolase peptide, all RA sera ($n = 28$) and HC sera ($n = 28$) were tested for reactivity to various citrullinated peptides in ELISA. Peptides from pro-filaggrin, proteoglycan, fibronectin, and EBNA-2 were tested for antibody reactivity.

As shown in Figure 3, significant RA antibody reactivity was found to the citrullinated pro-filaggrin ($p = 0.0119$), proteoglycan ($p = 0.0004$), fibronectin ($p = 0.0459$) and EBNA-2 ($p < 0.0001$) peptides compared to the HC. The proteoglycan and EBNA-2 peptides obtained the highest sensitivities of 50% and 68%, respectively. One HC serum showed low reactivity to EBNA-2 L (Figure 3b).

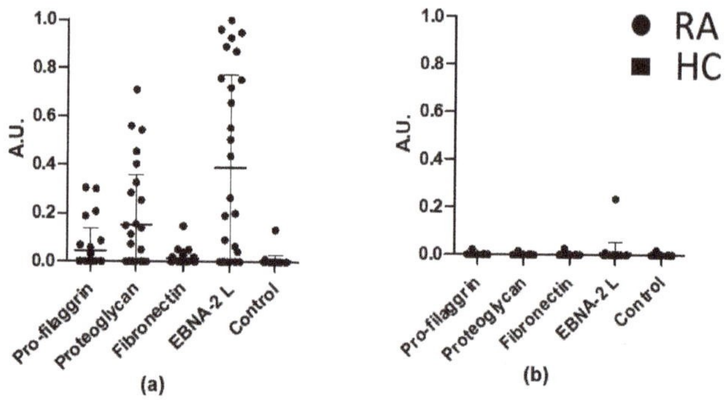


**Figure 3.** Reactivity of rheumatoid arthritis (RA) RA and healthy control (HC) sera to a selected peptide panel. (**a**) Screening of RA samples ($n$ = 28) to citrullinated peptides from pro-filaggrin, proteoglycan, fibronectin and EBNA-2. An Arg-containing pro-filaggrin peptide was used as control. (**b**) Screening of HC sera ($n$ = 28) to citrullinated peptides from pro-filaggrin, proteoglycan, fibronectin and EBNA2. An Arg-containing pro-filaggrin peptide was used as control. A.U. were defined as absorbances normalized relative to a positive RA control pool.

Thorough analysis of the reactivities of the RA sera showed that approximately 14% of the RA sera samples recognized all 4 peptides, whereas 18% reacted to 3 peptides. Note that approximately 54% of the samples recognized 1 or 2 peptides, whereas 14% of the RA cohort did not show reactivity to any of the peptides (Table 2, first column).

**Table 2.** Reactivities of the RA cohort to citrullinated peptides originating from pro-filaggrin, proteoglycan, fibronectin and EBNA-2.

| | Total RA Cohort | α-Enolase Positive RA Cohort | α-Enolase Negative RA Cohort |
|---|---|---|---|
| $n$ | 28 | 12 | 16 |
| 4 reactive peptides | 4 (14.3%) | 2 (16.7%) | 2 (12.5%) |
| 3 reactive peptides | 5 (17.9%) | 4 (33.3%) | 1 (6.3%) |
| 2 reactive peptides | 6 (21.4%) | 3 (25.0%) | 3 (18.8%) |
| 1 reactive peptide | 9 (32.1%) | 2 (16.7%) | 7 (43.8%) |
| 0 reactive peptides | 4 (14.3%) | 1 (8.3%) | 3 (18.8%) |

Of the 9 RA sera that only reacted with 1 peptide (Table 2, first column), 22% reacted with the proteoglycan peptide ($n$ = 2), 22% with the fibronectin peptide ($n$ = 2) and 56% ($n$ = 5) with EBNA-2 peptide.

When dividing the complete RA cohort into α-enolase-positive and -negative sera, it was observed that the RA samples in the α-enolase-positive cohort were prone to have a higher degree of overlapping antibody reactivities compared to the α-enolase-negative cohort. For instance, 50% of the RA sera in the α-enolase-positive cohort recognized 3 or 4 peptides compared to 19% in the α-enolase-negative cohort. Similarly, 42% of the α-enolase-positive cohort recognized 1 or 2 peptides compared to 63% for the α-enolase-negative cohort.

When examining the reactivity of the 3 cohorts (complete cohort, α-enolase-positive cohort, α-enolase-negative cohort) to the three most reactive peptides, pro-filaggrin, proteoglycan, and EBNA-2 L, similar results were obtained. As presented in Figure 4, significant overlapping reactivities were found between the 3 cohorts.

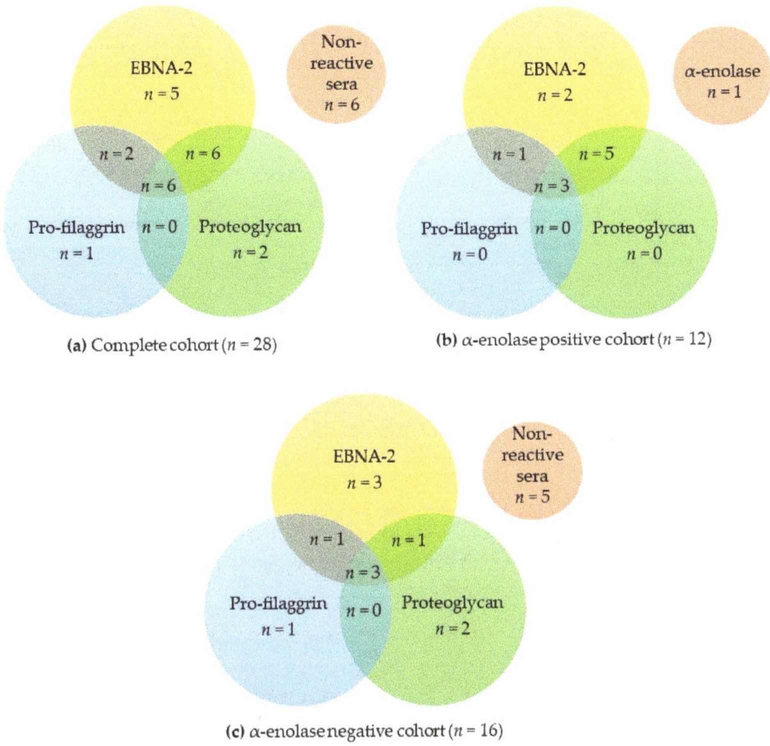


**Figure 4.** Venn diagram illustrating overlapping anti-citrullinated protein antibody reactivities to EBNA-2, pro-filaggrin and proteoglycan. (**a**) Reactivities of the complete rheumatoid arthritis (RA) cohort to the selected peptides ($n$ = 28). (**b**) Reactivities of α-enolase-positive RA serum samples ($n$ = 12). (**c**) Reactivities of α-enolase-negative RA serum samples ($n$ = 16).

Sera that only reacted to one peptide primarily recognized EBNA-2, as described earlier. Moreover, the most significant overlap in antibody reactivity was found between EBNA-2 and proteoglycan (Figure 4a,b), which is in accordance with the fact that these peptides obtained the highest sensitivities, as presented in Figure 3.

### *3.4. Reactivity of RA Sera to Linear and Cyclic Peptide Versions. Is the Effect of Cyclization on Antibody Sensitivity General?*

Previous findings indicated that the cyclic version of the α-enolase peptide had a higher sensitivity, although not statistically significant when compared to the linear peptide. To determine whether this effect is general or peptide-specific, RA reactivity to linear and cyclic citrullinated pro-filaggrin and EBNA-2 peptides were tested in ELISA.

Figure 5 illustrates the reactivity of RA and HC sera to the linear and cyclic Pro-filaggrin and EBNA-2 peptide. ACPA reactivity to the citrullinated peptides was significantly elevated compared to the HCs ($p = 0.0044$ for EBNA-2 L, $p < 0.0001$ for EBNA-2 C, $p = 0.0028$ for pro-filaggrin L, $p < 0.0001$ for pro-filaggrin C).

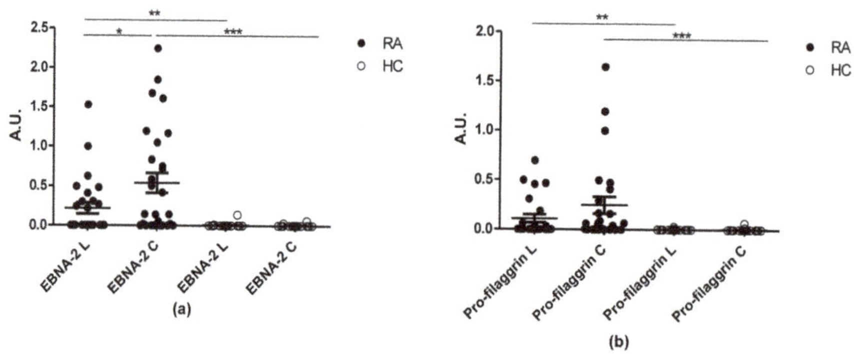


**Figure 5.** Reactivity of rheumatoid arthritis (RA) samples and healthy control (HC) samples to pro-filaggrin (25 RA and 25 HC samples) and EBNA-2 (28 RA and 28 HC samples) peptides in their linear and cyclic conformation. (**a**) Reactivity of RA and HC sera to citrullinated cyclic and linear EBNA-2 peptides. (**b**) Reactivity of RA and HC sera to citrullinated cyclic and linear pro-filaggrin peptides. A.U. were defined as absorbances normalized relative to a positive RA control pool. *** $p < 0.001$, ** $p < 0.01$, * $p < 0.05$.

Moreover, cyclization was observed to increase assay sensitivity, as the cyclic EBNA-2 and pro-filaggrin peptides had higher sensitivitities compared to the linear peptides ($p = 0.0331$), 67% and 44% of the RA sera reacted to the cyclic and linear EBNA-2 peptide, respectively, whereas 68% and 44% of the RA sera reacted to the cyclic and linear pro-filaggrin peptides, respectively. No statistically significant difference in antibody reactivity to the linear and cyclic pro-filaggrin peptides was found, although the antibody reactivity to the cyclic version was elevated (Figure 5b).

Collectively, these findings indicate that peptide conformation and peptide length affect antibody reactivity.

## 4. Discussion

In the present study, we analysed the reactivity of peptide-specific ACPAs to citrullinated epitopes and confirmed that factors such as peptide length and conformation notably influence antigen presentation.

The sensitivity of the full-length α-enolase peptide was determined to be approximately 44%, which is supported by earlier findings described in the literature [32]. ACPA reactivity to α-enolase was originally described by Lundberg et al., who showed that a cyclic peptide obtained the highest antibody reactivity among the human α-enolase and *Porphyromonas gingivalis* enolase peptides tested [32]. Similarly, no reactivity was found to the Arg-containing control peptide, confirming that the ACPAs are citrulline-specific (Figure 1). In addition to this, MS samples were tested for reactivity to the linear α-enolase peptide, as it has been described that citrullinated protein levels are elevated in MS [33], however, no reactivity was observed, confirming that ACPA reactivity is specific for RA.

The above mentioned experiments using α-enolase peptides were conducted in the absence of reducing agents, thus the linear peptides may in theory be able to cyclize during coating; however, as the effect of cyclization for α-enolase peptides was the same for pro-filaggrin and EBNA-2 peptides, where no cysteines were found in the linear peptides, we have reason to believe that the α-enolase peptides were found in a linear form during coating [23].

Screening of the truncated α-enolase peptides showed that all of the linear peptides (L-19-Cit, L-14-Cit, L-12-Cit and L-10-Cit) were roughly recognized to the same extent by the RA cohort, indicating that the length of the linear peptides is less important compared to the cyclic peptides. These findings confirm that the Cit–Gly motif in combination with a peptide backbone of approximately 10 amino acids, perhaps even shorter, is sufficient for antibody binding, as previously proposed [23]. In terms of reactivity, L-14-Cit, which is the second-longest peptide, showed the highest reactivity. In a study conducted using pro-filaggrin peptides, it was found that a linear 21-mer peptide and a linear 14-mer peptide were significantly recognized by ACPAs. Moreover, the 14-mer peptide obtained the highest sensitivity, which conforms to this study, favouring peptides of approximately 10–14 amino acids [27]. This effect may relate to the peptide structure. Even though a small number of amino acids are flanking the Cit–Gly motif in the shorter peptides, the peptides are still able to fold and thus acquire a specific conformation that appears to bind to ACPAs more efficiently. The exact reason remains to be determined.

Increased sensitivities were obtained for the longest (C-19-Cit) and the shortest (C-10-Cit) cyclic α-enolase peptides compared to the C-14-Cit and C-12-Cit peptides, indicating that peptide length and conformation are crucial for peptides containing 12–19 amino acids. These findings are supported by the literature, where it has been reported that the reduced number of amino acids in the cyclic structure may constrain the peptide in a more locked conformation, reducing the flexibility of the peptides and hence negatively influencing the ACPA reactivity [16]. The fact that the smaller peptide (C-10-Cit) was as sensitive as the longest peptide (C-19-Cit), and thus more sensitive than the mid-length peptides (C-12-Cit and C-14-Cit), is intriguing. This interesting reactivity pattern to truncated cyclic peptides could be further investigated by performing crystallography studies of the ACPA binding groove. Ultimately, these findings regarding truncated peptides (both linear and cyclic) confirm that peptide length and conformation are essential for antibody reactivity.

Concerning the aspect of cyclization, it has been reported that peptide cyclization has a positive effect upon antibody reactivity [14]. Our studies of pro-filaggrin and EBNA-2 cyclic and linear peptides revealed an evident increase in the antibody reactivity to the cyclic peptides compared to the linear peptides. These results are consistent with the longest α-enolase peptides (L-19-Cit and C-19-Cit). Nevertheless, for the α-enolase peptides, the reactivity to linear peptides did not depend on peptide length, whereas ACPA reactivity to cyclic peptides was length-dependent. This effect is in direct contrast to earlier findings using cyclic and linear pro-filaggrin peptides, where ACPA reactivity to both the linear and the cyclic peptides appeared to be length-dependent [27]. This may in theory be ascribed to the peptides used, as the pro-filaggrin and EBNA-2 peptides were biotinylated, whereas the α-enolase peptides had free terminals, hence the absence of biotin may have influenced the peptide coating. Thus, further analyses are necessary to confirm these results.

When focusing on the peptide sequence, a high degree of homology is found between pro-filaggrin and EBNA-2 in the C-terminal region, where positively charged and small amino acids are present (Table 1). The high degree of sequence homology may explain the similar sensitivities that the peptides yielded, suggesting that peptide sequence influences antibody reactivity. However, the Proteoglycan Cit peptide sequence does not have homology in the C-terminal end to pro-filaggrin and EBNA-2 and still yields a sensitivity of approximately 50%. Since several sequence patterns along with a Cit–Gly motif can be found among the peptides tested, these observations led to the hypothesis that a structural homology could be shared by the citrullinated peptides recognized by ACPAs. This remains to be elaborated.

Screening of the RA and the HC cohorts on a peptide panel, including four citrullinated peptides from pro-filaggrin, proteoglycan, fibronectin and EBNA-2, revealed a significantly different reactivity between RA and HC samples (Figure 3). The citrullinated peptides obtained the following sensitivities: Pro-filaggrin 32%, Fibronectin 36%, Proteoglycan 50% and EBNA-2 68%. In addition, more than 50% of the RA samples reacted with more than one peptide of the panel, confirming the ability of ACPAs to bind to several citrullinated targets. Further investigation analysing the overlapping reactivities of the RA sera within the peptide panel (Figure 3) showed that approximately 32% reacted with 3 or 4 peptides, 21% with 2 peptides, 14% did not react at all and, lastly, 32% of the cohort only showed reactivity to one peptide, which in 56% of the cases was the EBNA-2 peptide, suggesting the presence of EBNA-2-specific ACPAs. The fact that most of the RA samples interacted with more than one peptide supports the theory of the overlapping reactivity of ACPAs, highlighting the central role of the Cit–Gly motif together with the surrounding amino acids for antigen–antibody binding.

The previous findings led us to compare the ACPA reactivities between the peptide panel and the α-enolase peptides, dividing them into α-enolase-positive and -negative ones. As shown in Table 2, within the α-enolase-positive cohort ($n$ = 12), all the RA sera reacted with EBNA-2 peptide and no monospecific reactivity was detected to pro-filaggrin and proteoglycan peptides. One RA sample of the α-enolase-positive cohort only showed reactivity to the α-enolase peptide and not to the whole peptide panel, indicating that it was specific for the side-chains of the α-enolase epitope rather than the actual backbone in combination with a central Cit–Gly motif. In the α-enolase-negative cohort, more monospecific reactivities were observed. These findings are in accordance with the literature, which reports that approximately 15% of the RA sera are monospecific [30]. Additionally, although a small part of the RA sera reacted with only one citrullinated peptide, these results confirm the theory of overlapping and nonoverlapping ACPA reactivities. Here, the α-enolase-positive cohort was regarded to have overlapping ACPAs, which are considered as backbone-dependent, whereas the α-enolase-negative cohort was accounted to the nonoverlapping group of ACPAs, which depend on the flanking amino acid side-chains to establish a stable antigen–antibody interaction [16]. Collectively, the RA sera that were positive for α-enolase turned out to be more overlapping within the peptide panel compared to the α-enolase-negative cohort.

## 5. Conclusions

RA is an autoimmune disease that affects many people and reduces their quality of life. Thus, early diagnosis of the disease is fundamental to undertake therapy as soon as possible and to prevent disease progression. One of the most important diagnostic criteria is the detection of autoantibodies directed to a variety of citrullinated antigens in the serum of the patient. The origin of these autoantibodies is still unknown, and thus, gaining knowledge about the epitopes that ACPAs are able to recognize is important in the development of more sensitive and more specific diagnostic tools and to obtain a better understanding of the pathophysiological mechanisms. For these purposes and to elucidate the etiology of RA, interactions between ACPAs and potential candidate (auto)antigens were analysed.

Regarding the experiments to further investigate the structure and composition of ACPAs, these results confirm their cross-reactive nature. The analysed RA sera showed a different reactivity pattern to the citrullinated peptide panel in relation to the α-enolase-positive and -negative cohorts. This result indicates that ACPAs can be divided into two categories based on their ability to bind to a wider or a more limited number of citrullinated targets, which, respectively, reflects the peptide backbone and peptide side-chains dependencies. Especially, the ability of ACPAs to react with various citrullinated targets can be explained by a similar structure of citrullinated epitopes. Early studies showed that positively charged and small amino acids in the C-terminal end relative to citrulline yield high sensitivities. However, not all the peptides tested in this study have these features but can still interact with ACPAs (Table 1). Therefore, a structural homology, rather than sequence homology, should be important for ACPA recognition of the citrullinated targets, and this would, in addition, support the theory that the overlapping group of ACPAs is backbone-dependent. To obtain a better knowledge of the structure and to find a pattern that brings together all the citrullinated targets that interact with ACPAs, circular dichroism analyses could be performed.

Within the peptides tested on the RA cohort, the epitope originating from α-enolase, a known autoantigen in RA, was confirmed to have a lower sensitivity compared to available commercial assays and other citrullinated peptides, such as the EBNA-2 peptide, which was confirmed to be a highly sensitive substrate, even though contribution of EBV infection to RA onset still needs to be clarified.

Collectively, this study contributes to the understanding of the nature of ACPAs.

**Author Contributions:** Conceptualization, G.H., P.R.H., P.R. and N.H.T.; methodology, I.F., G.H., N.H.T. and P.R.; software, I.F. and N.H.T.; validation, I.F., G.H. and N.H.T.; formal analysis, I.F. and N.H.T.; investigation, I.F., G.H., N.H.T. and P.R.; resources, G.H., J.F. and P.R.; data curation, I.F. and N.H.T.; writing—original draft preparation, I.F. and N.H.T.; writing—review and editing, I.F., G.H., N.H.T., P.R.H. and P.R.; supervision, G.H., N.H.T. and P.R.; project administration, G.H., N.H.T., P.R. and P.R.H.; funding acquisition, G.H., N.H.T. and P.R. All authors have read and agreed to the published version of the manuscript.

**Funding:** This research was funded by the Lundbeck foundation, grant number R231-2016-3622.

**Informed Consent Statement:** Patient consent was waived as the samples were tested anonymously when tested for diagnostic purposes. The tested samples were not traceable.

**Conflicts of Interest:** The authors declare no conflict of interest.

## References

1. Scott, D.L.; Wolfe, F.; Huizinga, T.W. Rheumatoid arthritis. *Lancet* **2010**, *376*, 1094–1108. [CrossRef]
2. Aletaha, D.; Smolen, J.S. Diagnosis and Management of Rheumatoid Arthritis: A Review. *JAMA* **2018**, *320*, 1360–1372. [CrossRef]
3. Myasoedova, E.; Crowson, C.S.; Kremers, H.M.; Therneau, T.M.; Gabriel, S.E. Is the incidence of rheumatoid arthritis rising?: Results from Olmsted County, Minnesota, 1955–2007. *Arthritis Rheum.* **2010**, *62*, 1576–1582. [CrossRef]
4. Smolen, J.S.; Aletaha, D.; McInnes, I.B. Rheumatoid arthritis. *Lancet* **2016**, *388*, 2023–2038. [CrossRef]
5. Littlejohn, E.A.; Monrad, S.U. Early Diagnosis and Treatment of Rheumatoid Arthritis. *Prim. Care Clin. Off. Pract.* **2018**, *45*, 237–255. [CrossRef]
6. Listing, J.; Kekow, J.; Manger, B.; Burmester, G.-R.; Pattloch, D.; Zink, A.; Strangfeld, A. Mortality in rheumatoid arthritis: The impact of disease activity, treatment with glucocorticoids, TNF$\alpha$ inhibitors and rituximab. *Ann. Rheum. Dis.* **2013**, *74*, 415–421. [CrossRef]
7. Aletaha, D.; Neogi, T.; Silman, A.J.; Funovits, J.; Felson, D.T.; Bingham, C.O., III; Birnbaum, N.S.; Burmester, G.R.; Bykerk, V.P.; Cohen, M.D.; et al. 2010 Rheumatoid arthritis classification criteria: An American College of Rheumatology/European League Against Rheumatism collaborative initiative. *Arthritis Rheumatol.* **2010**, *62*, 2569–2581. [CrossRef]
8. Trier, N.H.; Holm, B.E.; Slot, O.; Locht, H.; Lindegaard, H.M.; Svendsen, A.; Nielsen, C.T.; Jacobsen, S.; Theander, E.; Houen, G. Application of synthetic peptides for detection of anti-citrullinated peptide antibodies. *Peptides* **2016**, *76*, 87–95. [CrossRef] [PubMed]
9. Van De Stadt, L.A.; De Koning, M.H.M.T.; Van De Stadt, R.J.; Wolbink, G.; Dijkmans, B.A.C.; Hamann, D.; Van Schaardenburg, D. Development of the anti-citrullinated protein antibody repertoire prior to the onset of rheumatoid arthritis. *Arthritis Rheum.* **2011**, *63*, 3226–3233. [CrossRef]
10. Ioan-Facsinay, A.; Willemze, A.; Robinson, D.B.; Peschken, C.; Markland, J.; van der Woude, D.; Elias, B.; Ménard, H.A.; Newkirk, M.; Fritzler, M.J.; et al. Marked differences in fine specificity and isotype usage of the anti-citrullinated protein antibody in health and disease. *Arthritis Rheum.* **2008**, *58*, 3000–3008. [CrossRef] [PubMed]
11. Klareskog, L.; Stolt, P.; Lundberg, K.; Källberg, H.; Bengtsson, C.; Grunewald, J.; Harris, H.E.; Ulfgren, A.-K.; Dahlqvist, S.R.; Eklund, A.; et al. A new model for an etiology of rheumatoid arthritis: Smoking may trigger HLA–DR (shared epitope)–restricted immune reactions to autoantigens modified by citrullination. *Arthritis Rheum.* **2005**, *54*, 38–46. [CrossRef] [PubMed]
12. Mil, A.V.D.H.-V.; Verpoort, K.N.; le Cessie, S.; Huizinga, T.W.J.; De Vries, R.R.P.; Toes, R.E.M. The HLA–DRB1 shared epitope alleles differ in the interaction with smoking and predisposition to antibodies to cyclic citrullinated peptide. *Arthritis Rheum.* **2007**, *56*, 425–432. [CrossRef]
13. Kallberg, H.; Padyukov, L.; Plenge, R.; Ronnelid, J.; Gregersen, P.; van der Helm-van Mil, E.; Toes, R.E.M.; Huizinga, T.W.; Klareskog, L.; Alfredsson, L. Epidemiological Investigation of Rheumatoid Arthritis study group; Gene-gene and gene-environment interactions involving HLA-DRB1, PTPN22, and smoking in two subsets of rheumatoid arthritis. *Am. J. Hum. Genet.* **2007**, *80*, 867–875. [CrossRef] [PubMed]
14. Wesoly, J.; Mil, A.V.D.H.-V.; Toes, R.; Chokkalingam, A.P.; Carlton, V.E.H.; Begovich, A.B.; Huizinga, T.W.J. Association of thePTPN22 C1858T single-nucleotide polymorphism with rheumatoid arthritis phenotypes in an inception cohort. *Arthritis Rheum.* **2005**, *52*, 2948–2950. [CrossRef]
15. Tarcsa, E.; Marekov, L.N.; Mei, G.; Melino, G.; Lee, S.-C.; Steinert, P.M. Protein Unfolding by Peptidylarginine Deiminase. *J. Biol. Chem.* **1996**, *271*, 30709–30716. [CrossRef]
16. Trier, N.H.; Houen, G. Epitope Specificity of Anti-Citrullinated Protein Antibodies. *Antibodies* **2017**, *6*, 5. [CrossRef]
17. Vossenaar, E.R.; Zendman, A.J.; van Venrooij, W.J.; Pruijn, G.J. PAD, a growing family of citrullinating enzymes: Genes, features and involvement in disease. *BioEssays* **2003**, *25*, 1106–1118. [CrossRef]
18. Aggarwal, R.; Liao, K.; Nair, R.; Ringold, S.; Costenbader, K.H. Anti-Citrullinated Peptide Antibody (ACPA) Assays and their Role in the Diagnosis of Rheumatoid Arthritis. *Arthritis Rheum.* **2009**, *61*, 1472–1483. [CrossRef]
19. Schellekens, G.A.; De Jong, B.A.; Hoogen, F.H.V.D.; Van De Putte, L.B.; Van Venrooij, W.J. Citrulline is an essential constituent of antigenic determinants recognized by rheumatoid arthritis-specific autoantibodies. *J. Clin. Investig.* **1998**, *101*, 273–281. [CrossRef]

20. Schellekens, G.A.; Visser, H.; De Jong, B.A.W.; Hoogen, F.H.J.V.D.; Hazes, J.M.W.; Breedveld, F.C.; Van Venrooij, W.J. The diagnostic properties of rheumatoid arthritis antibodies recognizing a cyclic citrullinated peptide. *Arthritis Rheum.* **2000**, *43*, 155–163. [CrossRef]
21. Trier, N.H.; Holm, B.E.; Heiden, J.; Slot, O.; Locht, H.; Lindegaard, H.; Svendsen, A.; Nielsen, C.T.; Jacobsen, S.; Theander, E.; et al. Antibodies to a strain-specific citrullinated Epstein-Barr virus peptide diagnoses rheumatoid arthritis. *Sci. Rep.* **2018**, *8*, 3684. [CrossRef]
22. Trier, N.H.; Holm, B.E.; Heiden, J.; Slot, O.; Locht, H.; Jensen, B.; Lindegaard, H.; Svendsen, A.; Nielsen, C.T.; Jacobsen, S.; et al. The use of synthetic peptides for detection of anti-citrullinated protein antibodies in rheumatoid arthritis. *J. Immunol. Methods* **2018**, *454*, 6–14. [CrossRef]
23. Trier, N.H.; Holm, B.E.; Slot, O.; Locht, H.; Lindegaard, H.M.; Svendsen, A.; Houen, G. Physical Characteristics of a Citrullinated Pro-Filaggrin Epitope Recognized by Anti-Citrullinated Protein Antibodies in Rheumatoid Arthritis Sera. *PLoS ONE* **2016**, *11*, e0168542. [CrossRef]
24. Dam, C.E.; Houen, G.; Trier, N.H. The dependency on neighboring amino acids for reactivity of anti-citrullinated protein antibodies to citrullinated proteins. *Scand. J. Clin. Lab. Investig.* **2016**, *76*, 417–425. [CrossRef] [PubMed]
25. Kinloch, A.; Tatzer, V.; Wait, R.; Peston, D.; Lundberg, K.; Donatien, P.; Moyes, D.; Taylor, P.C.; Venables, P.J. Identification of citrullinated α-enolase as a candidate autoantigen in rheumatoid arthritis. *Arthritis Res.* **2005**, *7*, R1421–R1429. [CrossRef]
26. Trier, N.H.; Dam, C.E.; Olsen, D.T.; Hansen, P.R.; Houen, G. Contribution of Peptide Backbone to Anti-Citrullinated Peptide Antibody Reactivity. *PLoS ONE* **2015**, *10*, e0144707. [CrossRef] [PubMed]
27. Trier, N.H.; Leth, M.; Hansen, P.R.; Houen, G. Cross-reactivity of a human IgG1 anticitrullinated fibrinogen monoclonal antibody to a citrullinated profilaggrin peptide. *Protein Sci.* **2012**, *21*, 1929–1941. [CrossRef] [PubMed]
28. Snir, O.; Widhe, M.; Von Spee, C.; Lindberg, J.; Padyukov, L.; Lundberg, K.; Engström, Å.; Venables, P.J.; Lundeberg, J.; Holmdahl, R.; et al. Multiple antibody reactivities to citrullinated antigens in sera from patients with rheumatoid arthritis: Association with HLA-DRB1 alleles. *Ann. Rheum. Dis.* **2008**, *68*, 736–743. [CrossRef]
29. Ioan-Facsinay, A.; El-Bannoudi, H.; Scherer, H.U.; van der Woude, D.; Ménard, H.A.; Lora, M.; Trouw, L.; Huizinga, T.W.J.; Toes, R. Anti-cyclic citrullinated peptide antibodies are a collection of anti-citrullinated protein antibodies and contain overlapping and non-overlapping reactivities. *Ann. Rheum. Dis.* **2010**, *70*, 188–193. [CrossRef]
30. Kinloch, A.; Lundberg, K.; Wait, R.; Wegner, N.; Lim, N.H.; Zendman, A.J.W.; Saxne, T.; Malmström, V.; Venables, P.J. Synovial fluid is a site of citrullination of autoantigens in inflammatory arthritis. *Arthritis Rheum.* **2008**, *58*, 2287–2295. [CrossRef]
31. Sun, J.; Zhang, Y.; Liu, L.; Liu, G. Diagnostic accuracy of combined tests of anticyclic citrullinated peptide antibody and rheumatoid factor for rheumatoid arthritis: A meta-analysis. *Clin. Exp. Rheumatol.* **2013**, *32*, 11–21. [PubMed]
32. Lundberg, K.; Kinloch, A.; Fisher, B.A.; Wegner, N.; Wait, R.; Charles, P.; Mikuls, T.R.; Venables, P.J. Antibodies to citrullinated α-enolase peptide 1 are specific for rheumatoid arthritis and cross-react with bacterial enolase. *Arthritis Rheum.* **2008**, *58*, 3009–3019. [CrossRef] [PubMed]
33. Moscarello, M.A.; Mastronardi, F.G.; Wood, D.D. The Role of Citrullinated Proteins Suggests a Novel Mechanism in the Pathogenesis of Multiple Sclerosis. *Neurochem. Res.* **2007**, *32*, 251–256. [CrossRef] [PubMed]

*antibodies* MDPI

*Article*

# Recombinant Antibody Production Using a Dual-Promoter Single Plasmid System

**Stefania C. Carrara [1,2,†], David Fiebig [1,2,†], Jan P. Bogen [1,2,†], Julius Grzeschik [2], Björn Hock [3] and Harald Kolmar [1,*]**

1 Institute for Organic Chemistry and Biochemistry, Technische Universität Darmstadt, Alarich-Weiss-Str. 4, D-64287 Darmstadt, Germany; stefania.carrara@ferring.com (S.C.C.); fiebig@biochemie-tud.de (D.F.); bogen@biochemie-tud.de (J.P.B.)
2 Ferring Darmstadt Laboratories, Alarich-Weiss-Str. 4, D-64287 Darmstadt, Germany; julius.grzeschik@ferring.com
3 Ferring International Center S.A, Chemin de la Vergognausaz 50, CH-1162 Saint Prex, Switzerland; bjorn.hock@ferring.com
* Correspondence: Kolmar@biochemie-TUD.de
† These authors contributed equally to this work.

**Abstract:** Monoclonal antibodies (mAbs) have demonstrated tremendous effects on the treatment of various disease indications and remain the fastest growing class of therapeutics. Production of recombinant antibodies is performed using mammalian expression systems to facilitate native antibody folding and post-translational modifications. Generally, mAb expression systems utilize co-transfection of heavy chain (*hc*) and light chain (*lc*) genes encoded on separate plasmids. In this study, we examine the production of two FDA-approved antibodies using a bidirectional (BiDi) vector encoding both *hc* and *lc* with mirrored promoter and enhancer elements on a single plasmid, by analysing the individual *hc* and *lc* mRNA expression levels and subsequent quantification of fully-folded IgGs on the protein level. From the assessment of different promoter combinations, we have developed a generic expression vector comprised of mirrored enhanced CMV (eCMV) promoters showing comparable mAb yields to a two-plasmid reference. This study paves the way to facilitate small-scale mAb production by transient cell transfection with a single vector in a cost- and time-efficient manner.

**Keywords:** monoclonal antibodies; promoters; bidirectional; antibody production; upstream processing

**Citation:** Carrara, S.C.; Fiebig, D.; Bogen, J.P.; Grzeschik, J.; Hock, B.; Kolmar, H. Recombinant Antibody Production Using a Dual-Promoter Single Plasmid System. *Antibodies* **2021**, *10*, 18. https://doi.org/10.3390/antib10020018

Academic Editor: Itai Benhar

Received: 24 March 2021
Accepted: 10 May 2021
Published: 13 May 2021



## 1. Introduction

With the growing interest in monoclonal antibodies (mAbs) for therapeutic applications, advances in antibody production have improved drastically over the last decades. Due to the more complex structure of antibodies, their production requires host cells capable of natively folding and modifying the mAb. Modifications include post-translational glycosylation, which is, among other functional properties, critical to reduce their immunogenicity [1]. For this purpose, mammalian cells fulfil the requirements as appropriate hosts for antibody production [2,3]. Advances in transfection protocols and cell engineering have boosted the use of suspension cell lines with the ability to grow at high densities, and increased production yields [4,5]. Further within the drug discovery and development process, stable cell lines are generated for the most promising candidate(s), while transient transfection is performed at earlier stages to yield research quantities of mAbs, sufficient for characterization of lead candidates. The accessibility of commercially available transfection reagents with high efficiencies and the use of disposable materials makes transient expression an efficient and cost-effective strategy during early drug discovery [6]. Human Embryonic Kidney 293 (HEK293) and Chinese Hamster Ovary (CHO) cells are commonly used for transient antibody expression, due to their high expression yields and human-like glycosylation patterns [1,7,8].

The basis of successful antibody production is the correct folding of individual chains, followed by their accurate assembly, resulting in functional heterotetrameric glycoproteins Misfolded or partially folded antibodies are degraded by the host cell's intrinsic quality control system, resulting in low production yields. Furthermore, antibodies with undesired folding are not able to effectively engage their target antigen or to mediate effector functions, have unfavourable pharmacokinetics, and tend to aggregate. Besides these biological limitations, purification of antibody products contaminated with aggregated or misfolded mAbs is a major hindrance in the downstream processing of therapeutic molecules and is currently the topic of numerous studies [9–12].

Antibody folding starts upon co-translational translocation into the endoplasmic reticulum (ER) [13]. Following the homodimerization of the two heavy chains (HC), the light chains (LC) are associated and covalently linked via disulphide bonds [14]. The glycosylation at Asn297 is linked to the CH2 backbone in a co-translational manner [15]. During translation, chaperones are involved to ensure the correct folding of the individual domains, as well as the final assembly of the tetrameric mAb [16].

At present, the largest part of transient production of mAbs is carried out using a two-plasmid system, also known as co-transfection, for the expression of *lc* and *hc*, with each gene driven by its own promoter and transcribed separately [17,18]. These are carried out, for the most part, with an equimolar ratio of heavy chain and light chain genes. Nonetheless, contradictory results have been reported using Expi293-F cells. While some publications report that an equimolar gene ratio results in the highest yield of fully assembled IgGs [8], others have described optimal expression with a 1:2 ratio of heavy and light chain genes, respectively [19].

A large drawback of the two-plasmid system is the moderate control over the relative expression of *lc* and *hc*, with fluctuating cell-to-cell transfection efficiencies [17]. On the other hand, a bidirectional (BiDi) vector with a dual-promoter can ensure the introduction of both *hc* and *lc* genes into each cell in equal amounts. However, for some applications it could be beneficial to have a stronger expression of one chain over the other. By choosing two suitable promoters controlling the transcription of *hc* and *lc*, different ratios can be achieved [20]. While diverse approaches have been developed throughout the years to advance and facilitate recombinant antibody production, the design of expression vectors plays a large role for optimization of expression yields. Recently, Bayat and colleagues (2018) compared the use of three different vector design strategies for the expression of IgG1 antibodies in CHO cells, namely using the conventional two-vector approach with *hc* and *lc* encoded separately, a bicistronic vector based on internal ribosome entry sites (IRES), and a dual-promoter single vector approach. All vectors were under the control of a human cytomegalovirus (CMV) promoter. Expression analysis revealed that the dual-promoter vector system resulted in the highest mAb yield [21].

Andersen and co-workers have previously shown the ability of the CMV enhancer to control two core CMV promoters simultaneously, resulting in efficient antibody expression [22]. With this basis, we sought to investigate different promoters in a bidirectional format to facilitate transient transfection, avoiding co-transfections. We also sought to simplify *hc* and *lc* gene cloning by establishing a one-step Golden Gate cloning procedure that relies on the simultaneous plasmid incorporation of *hc* and *lc* genes together with the bidirectional promoter sequence. By analysing *hc* and *lc* gene expression, as well as through subsequent quantification of fully-folded IgGs, promoter and enhancer element combinations were compared. Here, we show the use of a dual-promoter, single plasmid approach using divergent promoters for the transient expression of two FDA-approved antibodies, Durvalumab and Avelumab, using both Expi293-F and ExpiCHO-S cell systems. This work lays the foundation to facilitate small-scale mAb production in drug discovery programs in a more efficient manner.

## 2. Materials and Methods

### 2.1. Plasmids & Cloning of Constructs

To allow the individual exchange of different variable domains, the utilization of κ and λ isotypes, as well as the usage of different BiDi promoter combinations, the backbone of the mammalian destination (MD) vector was built in a cassette-like manner. For vector amplification in *E. coli*, a chloramphenicol resistance was utilized, adjacent to the colE1 and the f1 origins. A stuffer sequence, flanked by *Esp*3I restriction sites, was downstream of an inverse-orientated SV40 polyA sequence that was intended to be a terminator signal for the light chain cassette. Upstream of the stuffer sequence, a partial hinge followed by the CH2 and CH3 domains of a human IgG1 were encoded. Again, a SV40 polyA signal sequence served as a terminator signal (Figure 1A). The plasmid was de novo designed in silico and ordered at GeneArt (Regensburg, Germany).

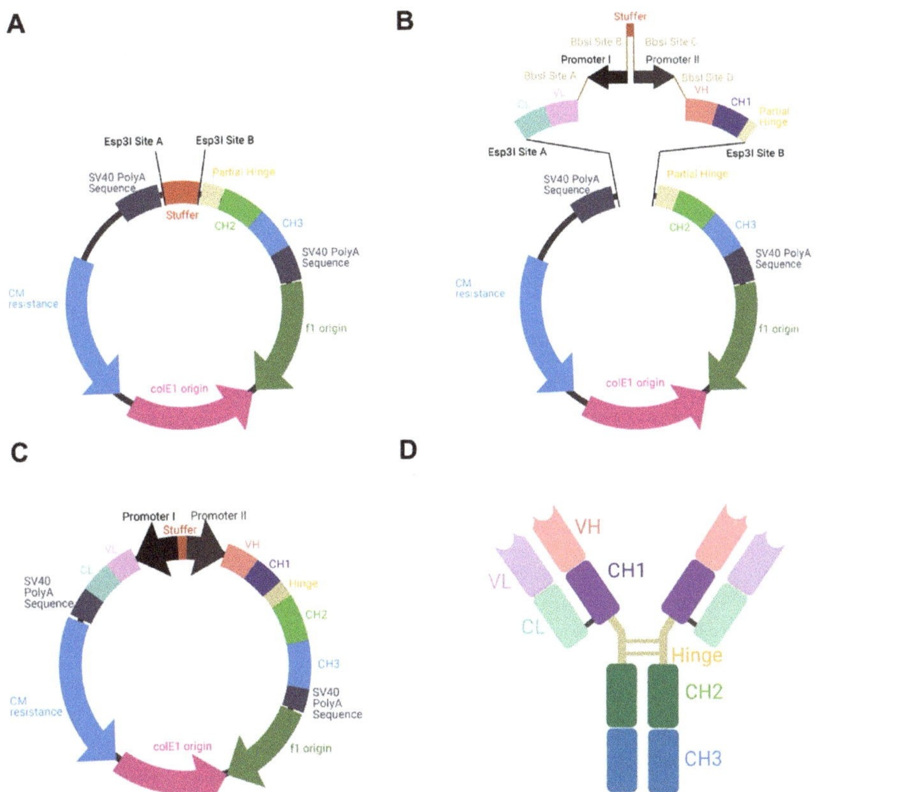


**Figure 1.** Schematic illustration of BiDi promoter system for antibody production. (**A**) The MD vector was designed to exhibit a 200-bp stuffer, flanked by *Esp*3I restriction sites (*Esp*3I sites A and B), adjacent to a SV40 polyA signal sequence and the regions encoding for hinge-CH2-CH3, terminated by a SV40 polyA signal sequence. (**B**) VL-CL and VH-CH1 amplicons can be inserted into MD by Golden Gate cloning utilizing *Esp*3I restriction sites (*Esp*3I sites A and B). The BiDi promoter can be chosen individually and is flanked by *Bbs*I sites (*Bbs*I sites A–D), compatible with the VL and VH sequences. (**C**) Golden Gate assembly results in a fully functional and re-circularized vector, with the light chain under the control of promoter I and the heavy chain under the control of promoter II. (**D**) Schematic representation of the resulting heterotetrameric IgG1 antibody using the same colour code as for the genetic elements.

The selected promoter sequences were either ordered as gene strings at Twist Bioscience (EF-1α, minCMV-enh-CMV (GenBank: MK764037) or PCR-amplified from the pTT5

CMV promoter cassette between bases 42–1185 (hereinafter referred to as eCMV) [23]. To allow for the correct orientation of the promoter sequences, individual primers were used to introduce the respective *Bbs*I Golden Gate cloning (GGC) signature overhangs. Genes for VH-CH1 and VL-CL of Durvalumab (Imfinzi, κ light chain) and Avelumab (Bavencio, λ light chain) were also ordered as gene strings, already bearing suitable signature sequences for *Esp*3I and *Bbs*I as well as their respective leader sequences. The 200-bp stuffer used between individual promoters consists of the non-functional 3′ coding region of the amp resistance gene for beta-lactamase, followed by ~40 bp of non-coding DNA. Assembly of the MD expression constructs was conducted with 75 ng destination vector and equimolar amounts of the respective fragments, 20 U *Bbs*I-HF, 10 U *Esp*3I, and 200 U T4-DNA ligase (NEB, Frankfurt, Germany) for 30 cycles (1 min; 16 °C; 37 °C). For the reference constructs, VH and VL genes were amplified incorporating *Sap*I restriction sites and then inserted into a pTT5-derived vector utilizing CH1-CH2-CH3 or κ/λ entry vectors using GGC as described before [24,25]. PCR reactions were performed utilizing Q5 polymerase (NEB) according to the manufacturer's protocol and purified using the Wizard SV Gel and PCR Clean-up System (Promega, Walldorf, Germany). All primers can be found in Table S1. The DNA sequence for the 2xeCMV BiDi construct can be found in Sequence S1 in the Supplementary Information.

*E. coli* XL1-blue were transformed utilizing the Golden Gate reaction mixtures and cultivated on chloramphenicol or ampicillin DYT agar plates for MD or pTT5 constructs, respectively. Resulting colonies were sequenced at MicroSynth SeqLab (Göttingen, Germany), and positive clones were utilized to inoculate 50 mL overnight cultures. Plasmid DNA for transient transfection was isolated using the PureYield Plasmid Midiprep System (Promega, Walldorf, Germany).

### 2.2. Cell Lines

Expi293-F and ExpiCHO-S cells were obtained from Thermo Fisher Scientific. Cells were incubated at 37 °C, 8% $CO_2$, 110 rpm, and sub-passaged every 3–4 days in their respective expression media, as described in the manufacturer's protocol (Thermo Fisher Scientific, Schwerte, Germany). Cell count and viability were measured using an automated cell counter (Bio-Rad TC-20) based on trypan blue staining. Cell densities were maintained between 0.3–4 × $10^6$ cells/mL and 0.2–6 × $10^6$ cells/mL for Expi293-F and ExpiCHO-S, respectively.

### 2.3. 24-Well Transfection

For gene expression and protein quantification, small-scale transfections using Axygen 24-well deep-well plates (Corning, New York, NY, USA) were performed. One day prior to transfection, cells were seeded into wells at a final cell density of 1.8 × $10^6$ or 3 × $10^6$ viable cells/mL in 2.5 mL expression medium for Expi293-F or ExpiCHO-S, respectively, and incubated under shaking conditions in a humified atmosphere at 37 °C, 8% $CO_2$, 225 rpm. The following day, the cell density was adjusted to 3 × $10^6$ or 6 × $10^6$ viable cells/mL in 2.5 mL expression medium for Expi293-F or ExpiCHO-S, respectively. DNA:Expifectamine complexes were incubated at room temperature with either 3 µg BiDi plasmid or 2 µg heavy and 2 µg light chain plasmid for co-transfections (two-plasmid reference) for 20 or 1 min for Expi293-F or ExpiCHO-S, respectively, before adding dropwise to the cells. Feeding procedures were carried out according to manufacturer's instructions. For gene expression analysis, cells were harvested 3 days post-transfection, while protein quantification was carried out 6 days post-transfection.

### 2.4. RNA Isolation

Three days post-transfection, Expi293-F or ExpiCHO-S cells were harvested by centrifugation and cell pellets were processed through a QIAshredder column (QIAGEN, Hilden, Germany). Total RNA extraction was carried out using RNeasy Mini Kit (QIAGEN) following the manufacturer's instructions. RNA concentration was determined spectroscopically using a NanoDrop One (Thermo Fisher), ensuring pure RNA was isolated with a A260/280 ratio of 2.0.

*2.5. Gene Expression Analysis by Reverse Transcription Quantitative Polymerase Chain Reaction (RT-qPCR)*

Expression levels of heavy and light chain (both κ and λ) were analysed using 100 ng RNA per well in Hard-Shell® 96-well PCR plates (Bio-Rad, Hercules, CA, USA) and iTaq Universal SYBR Green One-step Kit (Bio-Rad) with designed SYBR Green primers (Sigma Aldrich, Munich, Germany) using a CFX96 qPCR instrument (Bio-Rad). Relative expression levels were analysed using the integrated software from Bio-Rad (CFX Manager, Hercules, CA, USA) and normalized to housekeeping genes GAPDH and RPLP0 (IDT, Coralville, IA, USA). The primers used can be found in Table S2.

*2.6. Protein Purification*

To purify the antibodies from small-scale transfections, cells were harvested by centrifugation and cell culture supernatants were purified using Protein A HP SpinTrap columns (Cytiva, Freiburg im Breisgau, Germany) following the manufacturer's protocol. Antibodies were eluted in 0.1 M glycine-HCl, pH 2.7. Protein concentration was determined using a NanoDrop One (Thermo Fisher) using the corresponding molecular weights and extinction coefficients.

*2.7. Protein Quantification and Affinity Determination Using Biolayer Interferometry (BLI)*

Six days post-transfection, cells were harvested by centrifugation and the cell culture supernatants were sterile-filtered (0.45 µm). BLI experiments were performed on an Octet Red96 (FortéBio, Fremont, CA, USA). Using Protein A biosensors (Sartorius, Göttingen, Germany) for quantification, mAb concentration was measured from the cell culture supernatants. An in-house produced mAb was used as a standard within the range of 3.13–400 µg/mL.

For affinity determination, anti-human Fab-CH1 2nd generation (FAB2G) biosensors (Sartorius) were used. Purified antibodies were loaded onto the tips at 10 µg/mL until a layer thickness of 1 nm was reached. Association was measured using a serial dilution of His-PD-L1-TwinStrep (produced in-house). Kinetics were determined using Savitzky-Golay filtering and a 1:1 Langmuir binding model.

## 3. Results

To produce full-length antibodies in a bidirectional manner, we first designed the respective vector in silico. This vector encoded the fragment crystallizable (Fc) region of an IgG1, which is the most common isotype found in therapeutic antibodies [26]. In order to allow the flexible use for a variety of binders, the fragment antigen binding (Fab), which can be of the κ or λ isotype, was not encoded on the plasmid. Instead, a stuffer sequence, flanked by *Esp*3I sites was incorporated (Figure 1A).

As a reference antibody to establish different promoter combinations, the Fab of Durvalumab, an FDA-approved anti-PD-L1 antibody of the κ type, was chosen [27]. PCR amplicons of VH-CH1 and VL-CL were generated introducing *Esp*3I and *Bbs*I restriction sites. By utilization of a BiDi promoter system, flanked by *Bbs*I sites, a Golden Gate reaction resulted in a re-circularized vector (Figure 1B,C). In this process, the stuffer that was included in the parental MD vector was replaced by a CL-VL-PromoterI-Stuffer-PromoterII-VH-CH1 sequence. Owing to this cloning strategy, a straightforward exchange of different Fabs and different BiDi promoters is feasible. The resulting vector exhibited the functional ORFs for the heavy and the light chains, resulting in the production of full-length antibodies (Figure 1D).

### *3.1. Cloning Promoter Combinations*

Based on the findings from Andersen et al. [22] that a single enhancer adjacent to a bidirectional CMV promoter allows for efficient antibody production using a single expression plasmid, six different bidirectional promoter complexes were generated comprising not only the minCMV-CMV cassette, but also combinations of other strong promoters including their individual enhancer elements, such as the CMV with its major immediate early enhancer (MIE), the optimized CMV cassette from pTT5 (denoted as eCMV in this study), and the human translation elongation factor 1 alpha (EF-1α) promoter, that were selected based on their capability to produce fully-folded IgG molecules.

The MIE-CMV promoter, being one of the most commonly used in mammalian expression vectors, is designated as a strong driver for recombinant protein expression [28,29]. As shown before by Andersen and colleagues [22], the MIE enhancer is also capable of facilitating elevated expression levels in the divergently oriented minCMV promoter, although to a lesser extent. This was based on the presumption that the formation of the large transcription complex might be sterically hindered. As a reference, we selected a similar bidirectional promoter setup lacking the unique sequence upstream of the enhancer. Additionally, we went for a mirror-symmetric approach comprised of two individual CMV promoters, each having adjacent MIE enhancers that were separated by a 200-bp-stuffer. The eCMV cassette comprises—besides the MIE enhancer and core promoter—several additional regulatory elements that have been described to increase expression levels. These elements include the non-coding adenoviral tripartite leader sequence (TPL), the adenovirus major late promoter enhancer (MLP enh.), as well as distinct splicing sites allowing for prolonged mRNA stability [30,31]. Furthermore, we also utilized the strong human translation elongation factor 1 alpha promoter (EF-1α) that has proven to be advantageous over the CMV promoter in some cell types and in the expression of distinct proteins of interests [32,33].

Based on the modular setup of our MD vector and utilization of GGC, we generated different combinations of the aforementioned promoters to analyse for highest full-length antibody expression levels and product yield of Durvalumab. A schematic overview of the BiDi combinations is depicted in Figure 2.

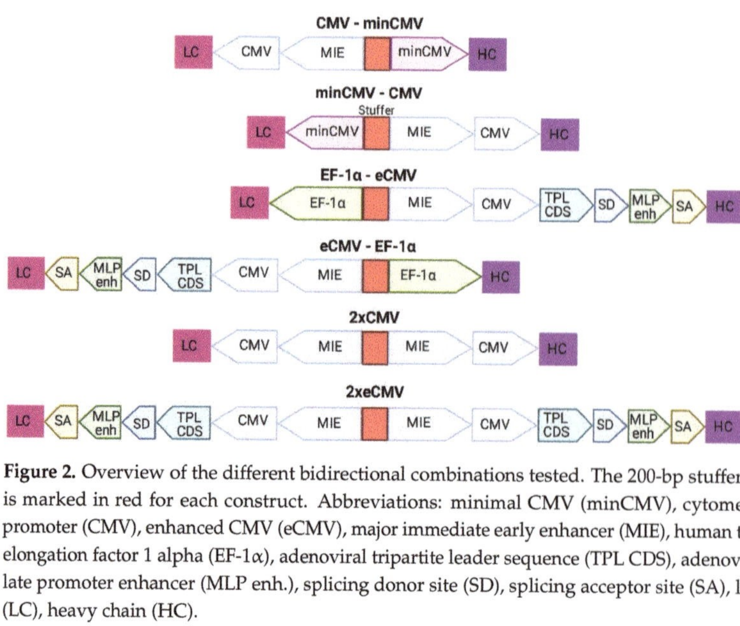


**Figure 2.** Overview of the different bidirectional combinations tested. The 200-bp stuffer sequence is marked in red for each construct. Abbreviations: minimal CMV (minCMV), cytomegalovirus promoter (CMV), enhanced CMV (eCMV), major immediate early enhancer (MIE), human translation elongation factor 1 alpha (EF-1α), adenoviral tripartite leader sequence (TPL CDS), adenovirus major late promoter enhancer (MLP enh.), splicing donor site (SD), splicing acceptor site (SA), light chain (LC), heavy chain (HC).

*3.2. Gene Expression Analysis in Mammalian Cells*

Transient transfections of all six promoter combinations were performed in 24-well plates using Expi293-F cells, an established cell line for transient antibody production. Three days post-transfection, cells were harvested, and RNA was isolated. Relative gene expression of *hc* and *lc* mRNA levels was measured by RT-qPCR (Figure 3).

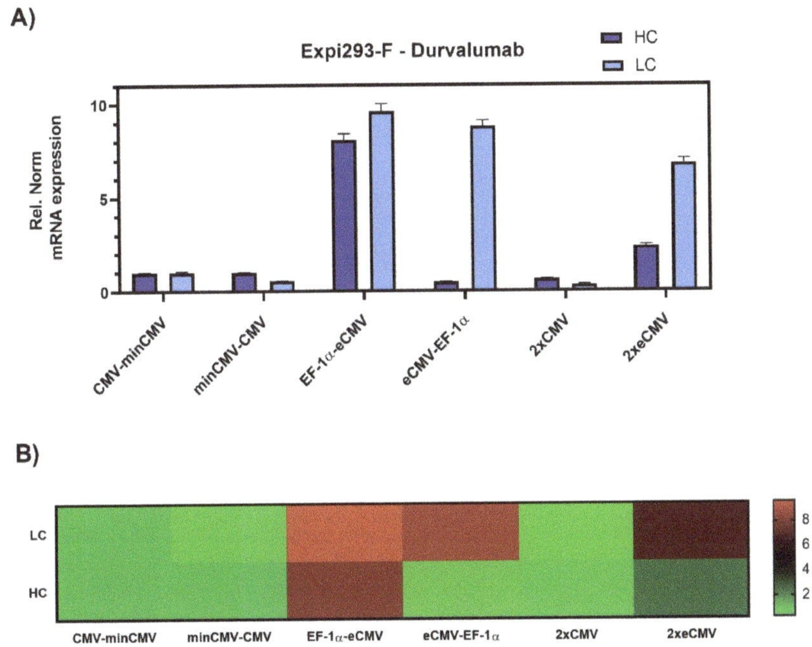


**Figure 3.** Gene expression analysis of heavy and light chain genes after transient transfection of Durvalumab in Expi293-F cells. (**A**) Bar chart representing heavy (dark blue) and light (light blue) chain mRNA expression in the different constructs. Values are relative to the CMV-minCMV construct and normalised to housekeeping genes GAPDH and RPLP0. Error bars represent the standard error of the mean of technical triplicates. (**B**) Heat map representation of gene expression analysis. The relative normalised gene expression for light and heavy chain mRNA is shown on the right.

The data was set relative to the CMV-minCMV construct and normalised to housekeeping genes. Looking at relative mRNA levels, both variants with minCMV and CMV, independent of the promoter orientation, did not yield high mRNA expression for either *lc* or *hc*. Remarkably, the combinations with the EF-1α promoter and the enhanced CMV cassette (eCMV) showed significant differences in expression levels depending on their orientation. Steering *lc* expression with the EF-1α promoter (EF-1α-eCMV) resulted in a 9.5-fold upregulation in *lc* and 8-fold upregulation in *hc* mRNA levels. Conversely, having the more potent eCMV in the light chain direction and EF-1α in the heavy chain direction (eCMV-EF-1α) led to a 9-fold *lc* upregulation, and low relative *hc* expression levels. Based on these results, mirrored constructs containing both promoter and enhancer cassettes in both directions were tested, one with the traditional CMV promoter (2xCMV), and the other with the eCMV cassette (2xeCMV). Interestingly, the 2xCMV construct did not result in increased *lc* or *hc* mRNA levels compared to the bidirectional construct containing two identical eCMV promoters or constructs containing two different promoters. The BiDi combination of two mirrored eCMV promoter cassettes (2xeCMV) yielded in a 7-fold upregulation of *lc* mRNA and a modest upregulation in *hc* levels.

As promoter strength may vary depending on the cell line, particularly for the CMV promoter [33], ExpiCHO-S cells were also investigated. Gene expression analysis resulted in similar results as in Expi293-F cells, with the 2xeCMV complex showing the strongest upregulation in both *hc* and *lc* mRNA levels compared to the other promoter combination, namely a modest 10-fold upregulation of *hc*, and a 20-fold upregulation of *lc* mRNA (Figure 4).

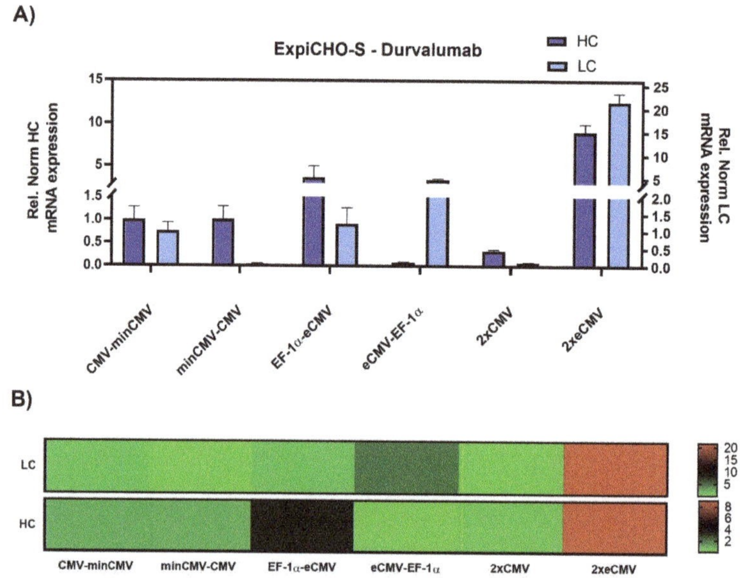


**Figure 4.** Gene expression analysis of heavy and light chain genes after transient transfection of Durvalumab in ExpiCHO-S cells. (**A**) Bar chart representing heavy (dark blue) and light (light blue) chain mRNA expression in the different constructs. Values are relative to the CMV-minCMV construct and normalised to housekeeping genes GAPDH and RPLP0. Error bars represent the standard error of the mean of technical triplicates. (**B**) Heat map representation of gene expression analysis. The relative normalised gene expression for light and heavy chain mRNA is shown on the right with their respective scales.

## 3.3. *Protein Yield Determination via BLI*

As mRNA transcript levels do not indicate successful secretion of fully functional recombinant antibodies, protein quantification studies were performed. Expi293-F cells were transiently transfected and the amount of secreted therapeutic antibody Durvalumab [27] was quantified by biolayer interferometry (BLI) using sterile-filtered cell culture supernatants six days post-transfection. The mAb concentrations for the different constructs were interpolated from a standard curve generated using an in-house produced mAb. In line with the gene expression analysis, protein quantification resulted in a clear ranking of the different promoter combinations (Figure 5).

A)

| Construct | mAb concentration (µg/mL) | Rank |
|---|---|---|
| 2xeCMV | 353.10 | 1 |
| EF1a-CMV | 42.90 | 2 |
| CMV-EF1a | 10.30 | 3 |
| 2xCMV | 10.30 | 3 |
| CMV-minCMV | 3.99 | 4 |
| minCMV-CMV | Too low | 5 |

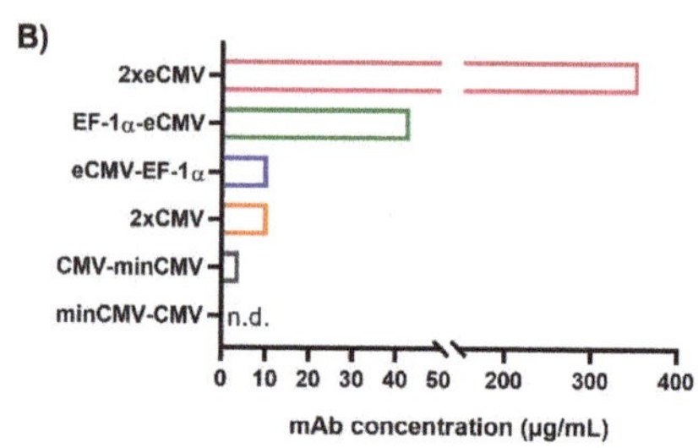


**Figure 5.** Protein quantification of Durvalumab in cell culture supernatants from transfected Expi293-F cells. (**A**) Table showing the mAb concentrations from 24-well transient transfections, listed according to their rank. The ranks 1–5 were set based on the mAb concentration of the different BiDi combinations. (**B**) Bar chart representation of BiDi mAb concentrations.

From the bidirectional combinations tested with transient transfections, the 2xeCMV BiDi construct showed the highest mAb concentration with 353 μg/mL (Figure 5). Interestingly, mRNA expression was higher for EF-1α-eCMV compared to 2xeCMV for both *hc* and *lc*, but antibody yield was significantly reduced in Expi293-F cells (Figure 3). This may likely be due to the fact that EF-1α-eCMV displayed similar *hc* and *lc* mRNA expression patterns, while 2xeCMV revealed a higher accumulation of the *lc* compared to the *hc* mRNA. It is well known that the expression of excess light chain over heavy chain is often beneficial for antibody production [34–37]. These findings corroborate that 2xeCMV resulted in the most promising bidirectional promoter combination, particularly in view of the fact that the usage of this promoter combination resulted, both in Expi293-F and ExpiCHO-S cell lines, in significantly enhanced mRNA synthesis with an excess of *lc* over *hc*.

### *3.4. Correlation of mRNA and Protein Levels*

After performing both mRNA and protein studies, the correlation of the cycle threshold (Ct) values from RT-qPCR and the mAb concentrations from protein quantification in Expi293-F was determined (Figure 6A). Overall, a Pearson's coefficient of −0.6407 was calculated, indicating, as expected, a negative correlation between Ct values and mAb concentration. Looking at the respective Ct values of either the *hc* or *lc* for 2xeCMV, one can observe they have a low Ct value and resulted in the largest mAb yield. While some claim that abundance of *hc* may hinder productivity, it appears that the excess of *lc* allowing for correct mAb folding and assembly is sufficient for higher mAb yields. On the contrary, the Ct values of eCMV-EF-1α *lc* is similar to that of 2xeCMV, with the only difference being in the *hc* expression, ultimately resulting in much lower yields. Thus, this further shows the importance of a fine-tuned *hc* and *lc* expression, substantiating the potential of 2xeCMV for a BiDi antibody production system.

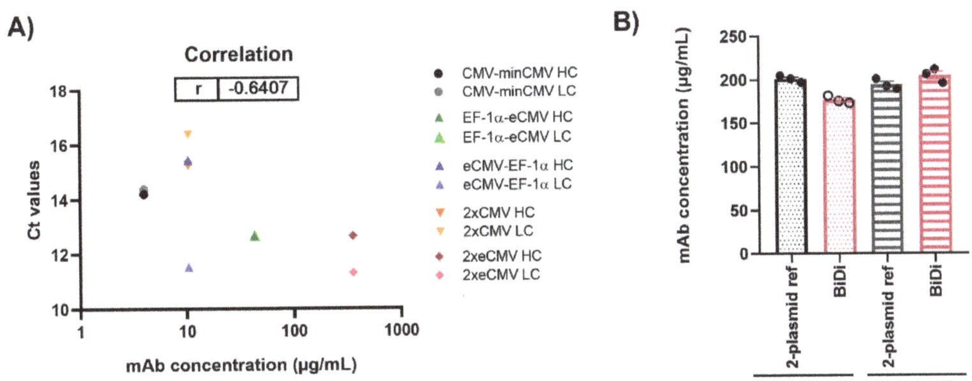


**Figure 6.** (**A**) Correlation of mAb concentration and Ct values for both heavy and light chain expression in Expi293-F for production of Durvalumab. (**B**) Quantification of antibody concentration for the production of Durvalumab and Avelumab using either a 2-plasmid reference or the BiDi 2xeCMV construct. Error bars represent the standard error of the mean of biological triplicates, while the symbols represent the individual measurements.

### *3.5. Antibody—And Light Chain-Independence*

To ensure the promoter combination used for our BiDi technology was not antibody- or light chain isotype-dependent, the FDA-approved anti-PD-L1 antibody Avelumab of the λ isotype was also tested [38]. As 70% of approved antibodies belong to the κ type [39,40], only the most promising 2xeCMV construct was used to validate the established system for a λ-based mAb, compared to the conventional co-transfection approach. For co-transfection, Expi293-F cells were transfected using a 1:1 ratio of HC and LC DNA, with each plasmid carrying the same promoter and enhancer cassette as the bidirectional vector, namely eCMV. As can be appreciated from Figure 6B, there was no significant variation in antibody yield between transfections with BiDi and the two-plasmid reference for either Durvalumab or Avelumab. Similarly, no variation was observed in ExpiCHO-S production (data not shown). Kinetics determination of both antibodies produced with either the two-plasmid reference or our established 2xeCMV BiDi plasmid bound to its target PD-L1 with comparable affinities (Table 1, Figures S1 and S2). Additionally, SDS-PAGE analysis under reducing conditions resulted in the expected heavy and light chains bands (Figure S3).

**Table 1.** Affinity determination of Avelumab and Durvalumab using either co-transfection or 2xeCMV BiDi plasmid.

| | Antibody | $K_D$ [pM] | |
|---|---|---|---|
| | | 2-Plasmid Ref. | BiDi |
| Expi293-F | Durvalumab | 594 | 364 |
| | Avelumab | 205 | 195 |
| ExpiCHO-S | Durvalumab | 562 | 522 |
| | Avelumab | 148 | 267 |

Thus, this indicated the established system is compatible with different binders and can be employed for both κ- and λ-light chain isotypes.

## 4. Discussion

This work describes the generation of a bidirectional vector construct for recombinant antibody expression using two eCMV promoter cassettes, controlling the expression of *lc* and *hc* individually in each direction. By performing a thorough analysis of different bidirectional promoter combinations with varying lengths and strengths, the 2xeCMV combination showed the most promising *hc* and *lc* mRNA synthesis in two regularly used mammalian cell lines, and, more importantly, the highest yields after protein quantification comparable to those using the conventional two-plasmid system. Other BiDi constructs showed potentially promising gene expression profiles, such as EF-1α-eCMV combination, with high relative mRNA levels of both *hc* and *lc* mRNA. Nonetheless, having higher levels of *hc* mRNA in the cells appear to curb productivity of fully folded IgG formation, resulting in drastically decreased antibody yields compared to both the two-plasmid reference and the 2xeCMV BiDi vector.

While using a bidirectional approach takes away some flexibility in terms of being able to alter the *hc:lc* ratios during co-transfection, the overexpression of both genes and, especially the excess *lc* expression, results in sufficient material for antibody hit screening. For convenience, we established a one-step cloning procedure for simultaneous plasmid incorporation of the heavy and light chain encoding segments obviating the need for generating two separate plasmids. Not only does this approach lower plasmid preparation efforts, but it also increases handling for transfection of numerous mAbs during screening and characterization as only a single plasmid is required. The use of two FDA-approved antibodies with either a κ- or λ-light chain shows there is no antibody- or light chain-dependence using this system, indicating that it can be implemented ubiquitously. Further options remain to increase the yields of IgG molecules, such as optimization of the stuffer region between the two eCMV promoter cassettes, potentially reducing any steric hinderance and increasing transcription efficiency [22,41].

In conclusion, this work displays the benefits of using a one-plasmid bidirectional system with 2xeCMV promoters for fully folded IgGs within drug discovery. In terms of practicality, handling of a single plasmid for antibody production may be superior to the conventional way. Moreover, yields of fully folded IgGs are comparable between the two systems. Future directions for this technology go beyond recombinant production of classical antibody formats, as reduction of the number of plasmid constructs could also be considered feasible for the expression of bispecific antibodies and other antibody formats in the frame of antibody drug discovery.

**Supplementary Materials:** The following are available online at https://www.mdpi.com/article/10.3390/antib10020018/s1, Table S1: Primers used for cloning of bidirectional constructs, Table S2: RT-qPCR primers for HC and LC constant regions, Figure S1: Affinity determination by BLI of antibodies produced in Expi293-F, Figure S2: Affinity determination by BLI of antibodies produced in ExpiCHO-S, Figure S3: SDS-PAGE analysis of purified antibodies, Sequence S1: DNA sequence of the designed Durvalumab-2xeCMV insert.

**Author Contributions:** Conceptualization, S.C.C., D.F., J.P.B., J.G. and H.K.; methodology, S.C.C., D.F., J.P.B.; investigation S.C.C., D.F. and J.P.B. data curation, S.C.C., D.F., J.P.B. and J.G. supervision B.H. and H.K. writing original draft, S.C.C., D.F. and J.P.B. writing review & editing S.C.C., D.F., J.P.B., J.G., B.H. and H.K. All authors have read and agreed to the published version of the manuscript.

**Funding:** This work was funded by the Ferring Darmstadt Labs at the Technical University of Darmstadt and by GPRD at Ferring Holding S.A., Saint Prex.

**Institutional Review Board Statement:** Not applicable.

**Informed Consent Statement:** Not applicable.

**Data Availability Statement:** The data presented in this study are available within this article and its Supplementary Materials.

**Acknowledgments:** S.C.C., D.F. and J.P.B. contributed equally to this work. The authors would like to thank GPRD for funding. The funders had no role in study design, data collection, data analysis, decision to publish, or preparation of the manuscript. We acknowledge support by the Deutsche Forschungsgemeinschaft (DFG—German Research Foundation) and the Open Access Publishing Fund of the Technical University of Darmstadt. Figures were created with Biorender and data was processed using GraphPad Prism 8.

**Conflicts of Interest:** J.G. and B.H. are employees of Ferring Pharmaceuticals, while S.C.C., D.F. and J.P.B. are employed by the Technische Universität Darmstadt in frame of a collaboration with Ferring Pharmaceuticals. All authors declare no conflicts of interest.

## References

1. Vink, T.; Oudshoorn-Dickmann, M.; Roza, M.; Reitsma, J.-J.; De Jong, R.N. A simple, robust and highly efficient transient ex-pression system for producing antibodies. *Methods* **2014**, *65*, 5–10. [CrossRef]
2. Carvalho, L.S.; Bravim da Silva, O.; Carneiro de Almeida, G.; Davies de Oliveira, J.; Parachin, N.S.; Carmo, T.S. Production processes for monoclonal antibodies. *Intech* **2017**. [CrossRef]
3. Li, F.; Vijayasankaran, N.; Shen, A.; Kiss, R.; Amanullah, A. Cell culture processes for monoclonal antibody production. *MAbs* **2010**, *2*, 466–479. [CrossRef] [PubMed]
4. Carrara, S.C.; Ulitzka, M.; Grzeschik, J.; Kornmann, H.; Hock, B.; Kolmar, H. From cell line development to the formulated drug product: The art of manufacturing therapeutic monoclonal antibodies. *Int. J. Pharm.* **2021**, *594*, 120164. [CrossRef]
5. Kunert, R.; Reinhart, D. Advances in recombinant antibody manufacturing. *Appl. Microbiol. Biotechnol.* **2016**, *100*, 3451–3461. [CrossRef] [PubMed]
6. Zhang, R.Y.; Shen, W.D. Monoclonal Antibody Expression in Mammalian. *Cells* **2012**, *907*, 341–358.
7. Graham, F.L.; Russell, W.C.; Smiley, J.; Nairn, R. Characteristics of a Human Cell Line Transformed by DNA from Human Adenovirus Type. *J. Gen. Virol.* **1977**, *36*, 59–72. [CrossRef]
8. Vazquez-Lombardi, R.; Nevoltris, D.; Luthra, A.; Schofield, P.; Zimmermann, C.; Christ, D. Transient expression of human an-tibodies in mammalian cells. *Nat. Protoc.* **2018**, *13*, 99–117. [CrossRef]
9. Gronemeyer, P.; Ditz, R.; Strube, J. Trends in Upstream and Downstream Process Development for Antibody Manufacturing. *Bioenineering* **2014**, *1*, 188–212. [CrossRef]
10. Shukla, A.A.; Hubbard, B.; Tressel, T.; Guhan, S.; Low, D. Downstream processing of monoclonal antibodies—Application of platform approaches. *J. Chromatogr. B* **2007**, *848*, 28–39. [CrossRef]
11. Van der Kant, R.; Karow-Zwick, A.R.; Van Durme, J.; Blech, M.; Gallardo, R.; Seeliger, D.; Rousseau, F. Prediction and reduction of the aggregation of monoclonal antibodies. *J. Mol. Biol.* **2017**, *429*, 1244–1261. [CrossRef]
12. Li, W.; Prabakaran, P.; Chen, W.; Zhu, Z.; Feng, Y.; Dimitrov, D.S. Antibody Aggregation: Insights from Sequence and Structure. *Antibodies* **2016**, *5*, 19. [CrossRef]
13. Bergman, L.W.; Kuehl, W.M. Formation of an intrachain disulfide bond on nascent immunoglobulin light chains. *J. Biol. Chem.* **1979**, *254*, 8869–8876. [CrossRef]
14. Baumal, R.; Potter, M.; Scharff, M.D. Synthesis, assembly, and secretion of gamma globulin by mouse myeloma cells. *J. Exp. Med.* **1971**, *134*, 1316–1334. [CrossRef] [PubMed]

15. Bergman, L.W.; Kuehl, W.M. Temporal relationship of translation and glycosylation of immunoglobulin heavy and light chains. *Biochemistry* **1978**, *17*, 5174–5180. [CrossRef]
16. Feige, M.J.; Hendershot, L.M.; Buchner, J. How antibodies fold. *Trends Biochem. Sci.* **2010**, *35*, 189–198. [CrossRef] [PubMed]
17. Ho, S.C.; Bardor, M.; Feng, H.; Mariati; Tong, Y.W.; Song, Z.; Yap, M.G.; Yang, Y. IRES-mediated Tricistronic vectors for enhancing generation of high monoclonal antibody expressing CHO cell lines. *J. Biotechnol.* **2012**, *157*, 130–139. [CrossRef]
18. Rita Costa, A.; Elisa Rodrigues, M.; Henriques, M.; Azeredo, J.; Oliveira, R. Guidelines to cell engineering for monoclonal antibody production. *Eur. J. Pharm. Biopharm.* **2010**, *74*, 127–138. [CrossRef] [PubMed]
19. Liu, C.Y.; Liu, J.; Yan, W.; Williston, K.; Irvin, K.; Chou, H.; Zmuda, J. Strategies for High-Titer Protein Expression Using the ExpiCHO and Expi293 Transient Expression Systems. Available online: https://assets.thermofisher.com/TFS-Assets/BID/posters/high-titer-protein-expression-expicho-expi293-poster.pdf (accessed on 5 March 2021).
20. Wodarczyk, C.; Reichenbächer, B.; Schulze, A.; Köhler, J.; Gerster, A.; Rehberger, B.; Müller, D. Increased antibody yield due to modification of LC and HC expression by gene regulatory elements. Available online: https://www.rentschler-biopharma.com/fileadmin/user_upload/Scientific-Posters/Rentschler_Poster__ESACT__2015.pdf (accessed on 17 February 2021).
21. Bayat, H.; Hossienzadeh, S.; Pourmaleki, E.; Ahani, R.; Rahimpour, A. Evaluation of different vector design strategies for the expression of recombinant monoclonal antibody in CHO cells. *Prep. Biochem. Biotechnol.* **2018**, *48*, 160–164. [CrossRef] [PubMed]
22. Andersen, C.R.; Nielsen, L.S.; Baer, A.; Tolstrup, A.B.; Weilguny, D. Efficient Expression from One CMV Enhancer Controlling Two Core Promoters. *Mol. Biotechnol.* **2010**, *48*, 128–137. [CrossRef]
23. Durocher, Y.; Perret, S.; Kamen, A. High-level and high-throughput recombinant protein production by transient transfection of suspension-growing human 293-EBNA1 cells. *Nucleic Acids Res* **2002**, *30*, e9. [CrossRef] [PubMed]
24. Bogen, J.P.; Carrara, S.C.; Fiebig, D.; Grzeschik, J.; Hock, B.; Kolmar, H. Expeditious Generation of Biparatopic Common Light Chain Antibodies via Chicken Immunization and Yeast Display Screening. *Front. Immunol.* **2020**, *11*. [CrossRef] [PubMed]
25. Bogen, J.P.; Storka, J.; Yanakieva, D.; Fiebig, D.; Grzeschik, J.; Hock, B.; Kolmar, H. Isolation of Common Light Chain Antibodies from Immunized Chickens Using Yeast Biopanning and Fluorescence-Activated Cell Sorting. *Biotechnol. J.* **2021**, *16*. [CrossRef] [PubMed]
26. Kretschmer, A.; Schwanbeck, R.; Valerius, T.; Rösner, T. Antibody Isotypes for Tumor Immunotherapy. *Transfus. Med. Hemotherapy* **2017**, *44*, 320–326. [CrossRef]
27. Alvarez-Argote, J.; Dasanu, C.A. Durvalumab in cancer medicine: A comprehensive review. *Expert Opin. Biol. Ther.* **2019**, *19*, 927–935. [CrossRef]
28. Boshart, M.; Weber, F.; Jahn, G.; Dorsch, -H.K.; Fleckenstein, B.; Schaffner, W. A very strong enhancer is located upstream of an immediate early gene of human cytomegalovirus. *Cell* **1985**, *41*, 521–530. [CrossRef]
29. Foecking, M.K.; Hofstetter, H. Powerful and versatile enhancer-promoter unit for mammalian expression vectors. *Gene* **1986**, *45*, 101–105. [CrossRef]
30. Ho, S.C.; Yap, M.G.; Yang, Y. Evaluating post-transcriptional regulatory elements for enhancing transient gene expression levels in CHO K1 and HEK293 cells. *Protein Expr. Purif.* **2010**, *69*, 9–15.
31. Sheay, W.; Nelson, S.; Martinez, I.; Chu, T.H.; Bhatia, S.; Dornburg, R. Downstream insertion of the adenovirus tripartite leader sequence enhances expression in universal eukaryotic vectors. *Biotechniques* **1993**, *15*, 856–862. [PubMed]
32. Suter, D.M.; Cartier, L.; Bettiol, E.; Tirefort, D.; Jaconi, M.E.; Dubois-Dauphin, M.; Krause, K.-H. Rapid Generation of Stable Transgenic Embryonic Stem Cell Lines Using Modular Lentivectors. *Stem Cells* **2006**, *24*, 615–623. [CrossRef]
33. Qin, J.Y.; Zhang, L.; Clift, K.L.; Hulur, I.; Xiang, A.P.; Ren, B.-Z.; Lahn, B.T. Systematic Comparison of Constitutive Promoters and the Doxycycline-Inducible Promoter. *PLoS ONE* **2010**, *5*, e10611. [CrossRef]
34. Schlatter, S.; Stansfield, S.H.; Dinnis, D.M.; Racher, A.J.; Birch, J.R.; James, D.C. On the Optimal Ratio of Heavy to Light Chain Genes for Efficient Recombinant Antibody Production by CHO Cells. *Biotechnol. Prog.* **2008**, *21*, 122–133. [CrossRef]
35. Gerster, A.; Wodarczyk, C.; Reichenbächer, B.; Köhler, J.; Schulze, A.; Krause, F.; Müller, D. A simple method to determine IgG light chain to heavy chain polypeptide ratios expressed by CHO cells. *Biotechnol. Lett.* **2016**, *38*, 2043–2049. [CrossRef]
36. Vanhove, M.; Usherwood, Y.-K.; Hendershot, L.M. Unassembled Ig Heavy Chains Do Not Cycle from BiP In Vivo but Require Light Chains to Trigger Their Release. *Immunity* **2001**, *15*, 105–114. [CrossRef]
37. Ho, S.C.; Koh, E.Y.; Van Beers, M.; Mueller, M.; Wan, C.; Teo, G.; Song, Z.; Tong, Y.W.; Bardor, M.; Yang, Y. Control of IgG LC:HC ratio in stably transfected CHO cells and study of the impact on expression, aggregation, glycosylation and conformational stability. *J. Biotechnol.* **2013**, *165*, 157–166. [CrossRef]
38. Powles, T.; Park, S.H.; Voog, E.; Caserta, C.; Valderrama, B.P.; Gurney, H.; Kalofonos, H.; Radulović, S.; Demey, W.; Ullén, A.; et al. Avelumab Maintenance Therapy for Advanced or Metastatic Urothelial Carcinoma. *N. Engl. J. Med.* **2020**, *383*, 1218–1230. [CrossRef]
39. Raybould, M.I.J.; Marks, C.; Krawczyk, K.; Taddese, B.; Nowak, J.; Lewis, A.P.; Bujotzek, A.; Shi, J.; Deane, C.M. Five computational developability guidelines for therapeutic antibody profiling. *Proc. Natl. Acad. Sci. USA* **2019**, *116*, 4025–4030. [CrossRef]
40. Grilo, A.L.; Mantalaris, A. The Increasingly Human and Profitable Monoclonal Antibody Market. *Trends Biotechnol.* **2019**, *37*, 9–16. [CrossRef]
41. Curtin, A.J.; Dane, A.P.; Swanson, A.E.; Alexander, I.; Ginn, S.L. Bidirectional promoter interference between two widely used internal heterologous promoters in a late-generation lentiviral construct. *Gene Ther.* **2008**, *15*, 384–390. [CrossRef]

*antibodies* MDPI

*Article*

# Characterization and Modeling of Reversible Antibody Self-Association Provide Insights into Behavior, Prediction, and Correction

**Carl Mieczkowski [1], Alan Cheng [2,*], Thierry Fischmann [3], Mark Hsieh [1], Jeanne Baker [1], Makiko Uchida [1], Gopalan Raghunathan [1], Corey Strickland [3] and Laurence Fayadat-Dilman [1]**

1 Discovery Biologics, Protein Sciences, Merck & Co., Inc., South San Francisco, CA 94080, USA; carl.mieczkowski@merck.com (C.M.); mark_hsieh@merck.com (M.H.); jeanne.baker@merck.com (J.B.); makiko.uchida@merck.com (M.U.); Raghu108@gmail.com (G.R.); laurence.fayadat-dilman@merck.com (L.F.-D.)
2 Discovery Chemistry, Modeling and Informatics, Merck & Co., Inc., South San Francisco, CA 94080, USA
3 Department of Chemistry, Modeling and Informatics, Merck & Co., Inc., Kenilworth, NJ 07033, USA; thierry.fischmann@merck.com (T.F.); corey.strickland@merck.com (C.S.)
* Correspondence: alan.cheng@merck.com; Tel.: +1-650-496-4834

**Abstract:** Reversible antibody self-association, while having major developability and therapeutic implications, is not fully understood or readily predictable and correctable. For a strongly self-associating humanized mAb variant, resulting in unacceptable viscosity, the monovalent affinity of self-interaction was measured in the low µM range, typical of many specific and biologically relevant protein–protein interactions. A face-to-face interaction model extending across both the heavy-chain (HC) and light-chain (LC) Complementary Determining Regions (CDRs) was apparent from biochemical and mutagenesis approaches as well as computational modeling. Light scattering experiments involving individual mAb, Fc, Fab, and Fab'2 domains revealed that Fabs self-interact to form dimers, while bivalent mAb/Fab'2 forms lead to significant oligomerization. Site-directed mutagenesis of aromatic residues identified by homology model patch analysis and self-docking dramatically affected self-association, demonstrating the utility of these predictive approaches, while revealing a highly specific and tunable nature of self-binding modulated by single point mutations. Mutagenesis at these same key HC/LC CDR positions that affect self-interaction also typically abolished target binding with notable exceptions, clearly demonstrating the difficulties yet possibility of correcting self-association through engineering. Clear correlations were also observed between different methods used to assess self-interaction, such as Dynamic Light Scattering (DLS) and Affinity-Capture Self-Interaction Nanoparticle Spectroscopy (AC-SINS). Our findings advance our understanding of therapeutic protein and antibody self-association and offer insights into its prediction, evaluation and corrective mitigation to aid therapeutic development.

**Keywords:** antibody; protein; self-association; self-interaction; developability; in silico prediction; computational modeling; viscosity; dynamic light scattering

**Citation:** Mieczkowski, C.; Cheng, A.; Fischmann, T.; Hsieh, M.; Baker, J.; Uchida, M.; Raghunathan, G.; Strickland, C.; Fayadat-Dilman, L. Characterization and Modeling of Reversible Antibody Self-Association Provide Insights into Behavior, Prediction, and Correction. *Antibodies* **2021**, *10*, 8. https://doi.org/10.3390/antib10010008

Received: 1 December 2020
Accepted: 1 February 2021
Published: 15 February 2021



## 1. Introduction

Monoclonal antibodies and biologics in general have enjoyed increasing success and utility as therapeutic agents addressing a variety of biological targets of interest. As of late 2019, 79 commercial monoclonal antibody or antibody-based therapeutics have been approved [1], with several hundred currently being evaluated in clinical development [2]. Central to a therapeutic antibody's selection and success is its developability profile, which is a key driver in pre-clinical and clinical lead nomination [3,4]. Previously, developability flags in therapeutic antibodies have been correlated to overall clinical success, clearly indicating that developability attributes may impact clinical development beyond drug product purity, stability and manufacturability [5]. The developability properties of therapeutic antibodies range from expression and purification amenability to its physicochemical stability and behavior, both in the drug product form and in vivo [6–8]. Other major developability properties, such as self-association, can directly impact manufacturability and formulation success [3,9], and even strongly correlate to non-specific binding and animal Pharmacokinetics and clearance, therefore, affecting its overall efficacy [10–13].

Therapeutic antibody self-association has been well studied from a rheological standpoint and is known to directly impact solution viscosity, injectability, and manufacturability [14,15]. Therapeutic antibody formulations as low as 13 mg/mL have been reported to appreciably self-associate, significantly increasing solution viscosity and decreasing solubility, precluding further development even at typical dose concentrations and formulation conditions [16]. Moreover, strong antibody self-interaction tends to manifest in high viscosities at higher formulation concentrations, such as 100 to 200 mg/mL (or approximately 0.7–1.4 mM for a typical monoclonal antibody) and beyond [3]. Increasing solution viscosity is due to a concentration-dependent oligomerization effect of self-interacting molecules, particularly antibodies, whereby effectively large polymeric structures give rise to dramatic changes in solution rheology [17,18]. A resulting increase in viscosity may be exponential, making process filtration and pumping operations difficult and infeasible, and in the drug product form, handling, injectability, and potentially even stability may be negatively impacted [3,19,20]. This behavior is a major negative developability attribute that is difficult to predict from sequence or structure and correct through molecular engineering and can halt further development and the selection of even the most promising large-molecule candidates [21].

High-concentration rheological behavior has significant importance in the selection of lead therapeutic candidates [22]. In general, molecular properties such as pI, net charge, and hydrophobicity can affect the rheology of antibody solutions [23]. Particularly at higher drug concentrations, it was shown that hydrophobic and charged surface patches result in increased self-interaction and solution viscosity above 100 mg/mL and approaching 200 mg/mL [24,25]. Such self-interactions can even lead to additional undesirable outcomes, such as opalescence, phase separation and gelling [26,27]. Antibody self-interactions have been reported to occur between antibody variable domains [21,28] as well as variable–constant interactions [16]. To possibly predict or correlate molecular properties to rheological outcomes, in silico computational approaches have been employed to ascertain a molecule's propensity to self-interact [29]. Surface behavior characteristics, such as zeta potential and net charge derived from modeling, have also been correlated to viscosity, [18,30,31] other self-interaction parameters, including AC-SINS and kD [32], and lead molecule selection and success [33,34]. In addition to the inherent or predicted properties of a molecule, temperature [35,36] and formulation conditions [9,28] can have a dramatic effect on solution rheology. Because of this, formulation approaches have been successful in mitigating self-association, such as modulating pH and ionic strength [37–39], or the addition of excipients, such as Arginine [40,41]. However, success is still dependent on the nature of the self-interacting therapeutic molecule and additional challenges and limitations are presented both from a process and formulation perspective.

In addition to directly measuring viscosity, a number of analytical methods and characteristics are informative of self-association, where typically size is directly measured or characterized. Notably, light scattering approaches, such as Dynamic Light Scattering (DLS), to determine the diffusion interaction parameter (kD) are informative and have been correlated to viscosity at higher concentrations [42]. Other measured or calculated characteristics, such as the second virial coefficient ($A_2$ or $B_{22}$), isoelectric point (pI) and zeta potential, have also been correlated to self-association phenomena [42–44]. Other techniques that evaluate self-interaction propensity or the propensity to interact with a column matrix involve Affinity-Capture Self-interaction Nanoparticle Spectroscopy (AC-SINS) [45–47], Analytical Ultracentrifugation (AUC) [48,49], Cross-Interaction Chromatography (CIC) and Standup Monolayer Adsorption Chromatography (SMAC) [43], many of which have been previously correlated to other self-interaction parameters, such as kD, $B_{22}$ and viscosity [39,50,51].

While self-association is impacted and defined by molecular surface properties, formulation effects, process considerations, and characterization approaches, the fundamental nature of an antibody's reversible self-interaction has yet to be fully gleaned, such as its typical orientation and strength. Additionally, while self-association can be routinely characterized and has been correlated to surface attributes elucidated both from crystal structures and computational models [16,21,29,52], predicting self-association from sequence or using in silico techniques to address and mitigate it is not routine or fully understood. Therefore, there is significant interest in understanding how to better predict and correct self-association at a sequence level to advance developability efforts. Previously, self-association of a recombinant monoclonal antibody was dramatically impacted and modulated by conservative mutagenesis of a single heavy-chain complementary determining region-3 (HC-CDR3) residue at position 104 [6], necessitating mitigation to find variants with lower, more acceptable self-association propensity. Developability attributes were evaluated for these variants, including DLS, AC-SINS, and viscosity, and correlations between these methods were apparent. However, this case study also offered a unique opportunity to further probe and to examine the intrinsic nature of self-interaction and how to better predict and correct it. Herein using similar variants, we verified viscosity profiles in a representative formulation buffer using the W104 and F104 mAb variants, validating the strongly self-associating F104 variant as a major developability risk to formulation efforts. We also mapped this strong self-association by evaluating the interaction of individual antibody domains using DLS, revealing this self-interaction was independent of the Fc domains and involved a blocking or likely face-to-face interaction between opposing Fab domains involving the CDR apparatus. We also determined the binding affinity of this strong self-interaction using both BIAcore Surface Plasmon Resonance (SPR) and Isothermal Titration Calorimetry (ITC), revealing, for the first time, the magnitude and binding affinity of representative, strong antibody self-interaction that arises in unacceptably high viscosity. We further evaluated in silico computational approaches, such as homology modeling and docking for their ability to predict and inform corrective engineering. Identified by homology modeling and docking were multiple key interacting residues in the HC and LC CDRs. These residues were subsequently and individually mutated, and all were found to dramatically affect self-interaction by DLS and AC-SINS. Given the highly specific and tunable nature of this self-interaction that structurally overlaps with its functionality, corrective engineering, while maintaining critical developability attributes such as target antigen binding, is challenging although demonstrated herein to be feasible. This study sheds much needed light on the nature of antibody self-association and how to potentially predict, correct and mitigate it.

## 2. Materials and Methods

### 2.1. Protein Expression and Purification

All recombinant antibodies (mAbs) and antibody binding fragments (Fabs) used are of the IgG1 isotype and were constructed by gene synthesis and expressed and purified as previously described [6]. Briefly, multi-liter (large) scale transient transfections were performed in 1 L shake flasks using the ExpiCHO Expression System (Thermo Fisher Scientific, Waltham, MA, USA) according to the manufacturer's protocol for all protein production. Antibodies were then affinity purified by Protein A MabSelect SuRe LX resin (GE Healthcare) in batch binding mode. Eluted antibody was then buffer exchanged into 20 mM Sodium Acetate pH 5.5 or 1X PBS pH 7.4 (Thermo Scientific). All purified recombinant antibodies were buffer exchanged overnight using 10 K MWCO Slide-A-Lyzer dialysis cassettes. Samples were concentrated using a Vivaspin™ ultrafiltration spin column with 10 K MWCO membrane (Sigma, Saint Louis, MO, USA). Concentration was determined by UV absorbance at 280 nm on a Nanodrop 2000 1-position Spectrophotometer (Thermo Scientific, Waltham, MA, USA).

For the preparation of Fab'2 constructs, the relevant mAb was digested by immobilized pepsin using the Pierce F(ab')2 Preparation Kit (Thermo Scientific, Waltham, MA, USA) according to the manufacturer protocol. Digested material was ProA affinity purified and the digested Fab'2 was recovered in the flow-through and buffer-exchanged, concentrated, and measured for concentration as performed for Fabs/mAbs.

All antibody variant identities were confirmed by intact LC-MS, as previously described [6].

### 2.2. Dynamic Light Scattering (DLS)

A DynaPro PlateReader II (Wyatt Technology, Santa Barbara, CA, USA) was used for all DLS experiments. First, 65 µL of recombinant antibody solution in 1X PBS pH 7.4 was added in duplicate to a 384-microwell glass bottom plate (Greiner bio-one, Austria). Then, 10 acquisitions were taken for a duration of 5 s at 25 °C using an auto-attenuated laser wavelength of 825 nm. Dynamics software version 7.8 (Wyatt Technology, Santa Barbara, CA, USA) was used for data analysis. All values, including hydrodynamic radius ($R_h$), diffusional coefficients (D), and diffusion interaction parameters (kD) are reported as the average of duplicate well collections. kD's in units of mL/g were obtained over the concentration range of 1–10 mg/mL at 1, 2, 4, 6, 8, and 10 mg/mL concentrations. kD was then calculated by the Dynamics software from plotting measured diffusional coefficients ($D_m$) versus sample concentration (C) and using the equation $D_m = D_o(1 + kD \cdot C)$ [53]. Negative kD's indicate self-interaction and increases in the magnitude indicate increased self-interaction. The $M_w$-R (or estimated molecular weight from radius) is calculated in Dynamics from measured diffusional coefficients using a standard spherical model.

### 2.3. Affinity-Capture Self-Interaction Nanoparticle Spectroscopy (AC-SINS)

Gold nanoparticles (Ted Pella) were exchanged into water and an 80/20 (*v*/*v*) capture antibody/non-capture antibody mixture (Jackson Immuno Research Labs, West Grove, PA, USA) was exchange into 20 mM Sodium Acetate pH 4.3. Then, 1 mL of gold nanoparticles were incubated overnight with 100 µL of antibody mixture. Gold nanoparticles were then pelleted by centrifugation and supernatant was decanted to achieve a final volume of 50 µL, followed by gentle mixing. Then, 5 µL of this concentrated nanoparticle suspension was added to 45 µL PBS solution containing 0.05 mg/mL of antibody of interest in a 384-well clear plate (Fisher Scientific, Waltham, MA, USA) and incubated at room temperature for 2 h in the dark. The plate was then quickly spun down at 3000 rpm and scanned from 450 to 650 nm using an EPOCH/2 Microplate reader from BioTek (Winooski, VT, USA). Values reported are averages of duplicate wells and are sample red shift wavelengths at maximum absorbance subtracting the blank reference (PBS only). Greater red shifts indicate increased self-interaction.

### 2.4. Viscosity Measurements

Viscosity was performed as previously described [6]. Briefly, dynamic viscosities were measured using a VROC initium (Rheosense, San Ramon, CA, USA) and processed using built-in software. Each reported viscosity value in Centipoise (cP) was the average of 10 measurements performed at 25 °C. For samples at 100 mg/mL, duplicate sample injections were performed.

### 2.5. Modeling

All modeling, including homology modeling, surface patch analysis, and protein–protein docking, was performed using MOE 2019.0102 (Chemical Computing Group, Montreal, QC, Canada), as described below. All calculations were performed using the Amber10:EHT force-field.

Homology models were produced for both F104 and W104 Fabs. The framework regions were modeled using the human antibody structure with highest sequence identity (PDB ID: 3sqo). Light chain CDR1, CDR2, and CDR3 were modeled using antibody fragment structures with PDB ID's of 5ken, 4 × 80, and 1 × 4, with sequence identities of 82%, 100%, and 78%, respectively. Heavy chain CDR1 and CDR2 were modeled using a structure with PDB ID of 5gs2, with sequence identities of 80% and 71%, respectively. Modeling of HC-CDR3 is known to be more challenging due to higher sequence variation and flexibility. Here, we used the three structural templates with highest sequence identity, 60% (PDB IDs of 1jgu, 1dbj, and 3ixt). The three homology models produced by MOE had similar loop conformations and the positioning of the heavy chain 104 residue, and so the model based on the 3ixt template, which has a Phe at the 104 position, was selected for the F104 Fab. For the W104 Fab, we used an identical approach but selected the 1dbj template for HC-CDR3 because it has a Trp at the heavy chain 104 position.

Surface patch analysis was performed in MOE on homology models of the F104 and W104 Fabs, using the default potential threshold value of 0.09 kcal/mol.

Protein–protein docking, as implemented in MOE, was used to model the self-interaction of the F104 Fab. The approach uses a coarse-grained representation for initial docking, followed by the refinement of docking poses using an all-atom representation. Docking was limited to the CDR regions, included side-chain flexibility, and used a docking potential that includes an extra term favoring the burial of hydrophobic patches identified by the surface patch analysis. This resulted in 94 docked models, which were then clustered by binding epitopes. The largest cluster identified by this method included 14 models and involved all four VL CDR residues and at least three of the five VH CDR residues of the largest hydrophobic patch. The 14 docked poses were visually inspected and found to cluster into four binding modes.

### 2.6. Isothermal Titration Calorimetry (ITC)

Isothermal titration calorimetry (ITC) was conducted using a MicroCal PEAQ-ITC Automated (Malvern Inc., Westborough, MA, USA) to determine the dissociation constant of antibody dimers. Fabs were prepared in PBS pH 7.4. The W104 Fab (371 μM) or F104 Fab (300 μM) in the ITC syringe was titrated into the matching buffer in the ITC cell at 25 °C. Reference power was set to 10 μcal/sec with initial delay of 60 s and stirring speed of 750 rpm. Injection volume was 0.4 μL for the first injection and 4 μL for subsequent injections and 13 total injections were made. Injection duration was 0.8 s for the first injection and 6 s for subsequent injections with 150 s spacing. Baseline was adjusted using buffer–buffer titration. Data analysis was done using MicroCal PEAQ-ITC Analysis Software using the dissociation model.

### 2.7. Surface Plasmon Resonance (BIAcore)

Binding affinity of Fab against captured mAb was determined by surface plasmon resonance (SPR) on a BIAcore T200 (Cytiva). The running buffer, 10 mM HEPES, 150 mM NaCl, 0.05% *v/v* Surfactant P20, 3 mM EDTA, pH 7.4 (HBS-EP+, Cytiva, Marlborough, MA, USA) was used for immobilization and reagent dilutions. All binding kinetics were measured at 25 °C.

For each injection cycle, mAbs were first captured in flow cells 2, 3 and 4 with an anti-human Fc antibody (Human Antibody Capture Kit, Cytiva, Marlborough, MA, USA) immobilized to the sensor chip (Series S CM5, Cytiva, Marlborough, MA, USA). Flow cell 1 with no captured mAb was used as a reference. Serial dilutions of the Fab, ranging in concentration from 1 to 32 μM, and buffer blanks were injected in multiple cycles over the captured mAbs and reference surfaces for a 60 s association followed by a 180 s dissociation. The surfaces were regenerated with a 30 s injection of 3 M $MgCl_2$ after each cycle.

Double referenced titration data were globally fit to a 1:1 Langmuir binding model to determine the association rate constant, ka (1/M·s), and the dissociation rate constant, kd (1/s), using the BIAcore T200 Evaluation Software version 2.0 (Cytiva, Marlborough, MA, USA). The equilibrium dissociation constant was calculated as KD (M) = kd/ka.

## 3. Results

### 3.1. *Viscosity Characterization for Two Antibody Variants*

Previously, a strongly self-associating mAb was reported where mutagenesis at a single position, HC-CDR3-104, dramatically affected the degree of self-interaction [6]. In Figure 1A, viscosity curves up to 100 mg/mL in a representative low pH and ionic strength formulation buffer (20 mM Sodium Acetate pH 5.5) are plotted for both the HC-W104 and HC-F104 variants. While the W104 variant has low viscosity up to 100 mg/mL (4.2 cP), the F104 variant increases exponentially up to 58.3 cP. At 100 mg/mL in PBS pH 7.4, viscosities were 4.8 and 43.7 cP, respectively, for the HC-W104 and HC-F104 variants (Figure 1B). These measured viscosities in both acetate and PBS buffers are over 2 times higher than what would be considered a typical allowable limit for either downstream processing or injectability [54]. Such strong and robust self-interaction rendered this F104 variant undevelopable, especially for a higher formulation concentration (>100 mg/mL). Since the HC-F104 variant had unacceptably high viscosity in both acetate and PBS buffers, further evaluation of the self-interaction herein focused on PBS formulations for all variants to remain consistent across all biophysical and analytical assays.

A

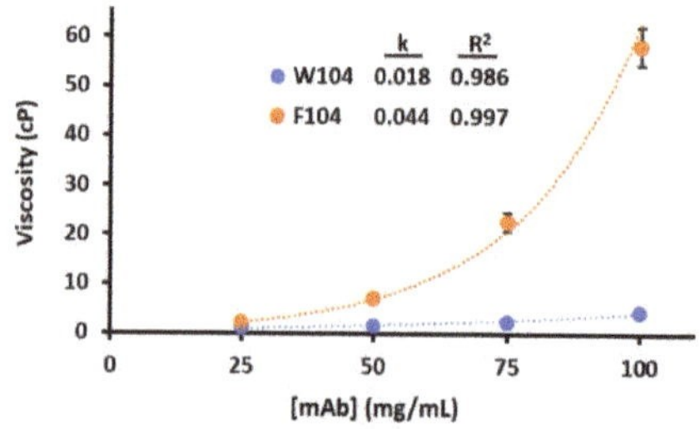


B

| Variant | Viscosity (cP) (Acetate) | *Viscosity (cP) (PBS) |
|---|---|---|
| W104 | 4.5 | 4.8 |
| F104 | 58.3 | 43.7 |

**Figure 1.** (**A**) Viscosity (dynamic) versus mAb concentration for W104 (blue circles) and F104 (orange circles) IgG1 mAb variants in 20 mM Sodium Acetate pH 5.5 from 25–100 mg/mL. Each viscosity curve was fit to the equation $y = Ae^{kx}$, where the exponential factor (k) and $R^2$ of the fit are given for each variant. (**B**) Table of viscosities for the two W104 and F104 mAb variants in both 20 mM Acetate pH 5.5 and PBS pH 7.4 at 100 mg/mL. * Previously reported [6].

### 3.2. *Biophysical Characterization and Modeling of Antibody Self-Association Using Individual Antibody Domains for Two Antibody Variants*

To better understand the nature of this self-association, individual domains consisting of Fab, Fab'2, and mAb, along with the conserved IgG1 Fc, were prepared for both F104 and W104 variants. In PBS pH 7.4, preparations ranging from 1–10 mg/mL of each construct and variant were prepared and evaluated by DLS. In Figure 2A, construct $M_W$ (based on sequence), kD obtained by DLS, and hydrodynamic radius ($R_h$) at both 1 and 10 mg/mL, are tabulated. As expected, the F104 mAb self-associated efficiently and a highly negative kD of −44.6 mL/g was obtained. By comparison, the HC-W104 mAb variant, while structurally similar to the HC-F104 mAb, has a significantly less negative kD of −15.3 mL/g. The F104 Fab'2 had a similar highly negative kD of −44.8 mL/g versus its mAb counterpart. This similarity alone is strong evidence that the Fab domains are self-interacting independent of the Fc domain, and further, in Figure 2B, a 1:1 mixture of IgG1 F104 Fab (kD = −9.5 mL/g) and IgG1-Fc (kD = −2.7 mL/g) does not increase self-interaction (or decrease kD), but rather an intermediate kD value results (−8.0 mL/g), supporting the idea that the Fab and Fc domains are not directly interacting.

**A**

| Molecule (IgG1) | Mw (kDa) | kD (mL/g) | Rh (nm) (1 mg/mL) | Rh (nm) (10 mg/mL) |
|---|---|---|---|---|
| W104 Fab | 48 | -4.5 | 3.5 | 3.7 |
| W104 Fab'2 | 98 | -18.7 | 5.7 | 6.9 |
| W104 mAb | 145 | -15.3 | 6.3 | 7.4 |
| F104 Fab | 48 | -9.5 | 3.9 | 4.3 |
| F104 Fab'2 | 98 | -44.8 | 8.4 | 14.4 |
| F104 mAb | 145 | -44.6 | 8.9 | 16.2 |

**B**

| Molecule (IgG1) | kD (mL/g) |
|---|---|
| Fc | -2.7 |
| Fab HC-W104 | -4.5 |
| Fab HC-F104 | -9.5 |
| Fab HC-W104 + Fc | -3.9 |
| Fab HC-F104 + Fc | -8.0 |

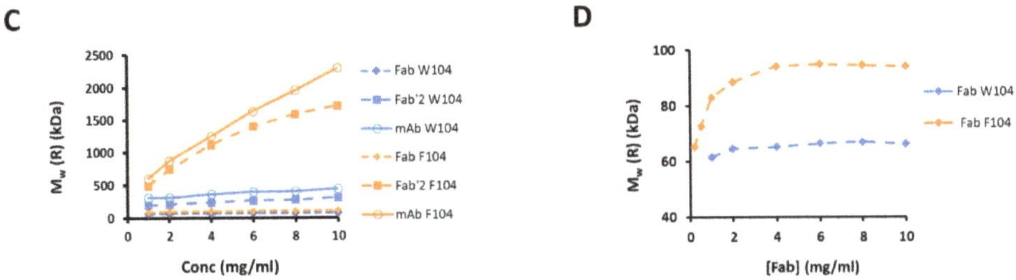


**Figure 2.** Dynamic Light Scattering (DLS) results for various constructs and HC-104 variants. (**A**) Tabulated molecular weight ($M_W$), known from sequence, and interaction parameter (kD) and hydrodynamic radius ($R_h$) obtained by DLS for both W104 and F104 variants in the Fab, Fab'2, and mAb forms. Highly negative kDs and large $R_h$ values are in red font. (**B**) kD tabulated for Fc and Fab constructs for W104 and F104, and Fab/Fc mixtures (1:1 molar ratio). (**C**) Calculated $M_w$ values derived from measured diffusional coefficients plotted against concentration for Fab, Fab'2, and mAb constructs for F104 (orange symbols) and W104 (blue symbols). Fab, Fab'2, and mAb are designated as diamonds, squares, and open circles, respectively. (**D**) Calculated $M_W$ values derived from measured diffusional coefficients for Fab forms only of W104 (blue diamonds) and F104 (orange squares).

Interestingly, the F104 variant, which has a high propensity to self-interact, has a much more negative kD in the Fab'2 and mAb forms than the Fab alone (−9.5 mL/g). Likewise, the highly negative kDs for Fab'2 and Mab constructs yield huge, reversible complexes by DLS, resulting in $R_h$ values of 14.4 and 16.2 nm at 10 mg/mL in PBS, respectively. Since the sizes of these complexes are not the result of aggregation driven by unfolding (aggregation measured by SE-UPLC is not significant, See Table S1) and are concentration dependent, these are reversible complexes formed through the participation of fully folded, native antibody or antibody domains consistent of typical self-interaction phenomena [28]. The sizes of the complexes formed at 10 mg/mL, a relatively low formulation concentration, for the F104 Fab'2 and mAb constructs are so large that molecular weights of 1714 and 2290 kDa (Figure 2C) are calculated from the $R_h$ values (assuming a standard spherical model), respectively. This corresponds to an approximate average size of the 17 non-covalently associated units for the Fab'2 construct and 16 non-covalently associated units for the mAb construct at 10 mg/mL. However, in the case of the F104 Fab construct, only a slightly negative kD of −9.5 mL/g is measured, and $R_h$ modestly increases from 3.9 nm to only 4.3 nm in the 1–10 mg/mL concentration range. At 6 mg/mL, a concentration where size appears to have already plateaued, a complex with a $M_w$ of 95 kDa is calculated, which approximates a Fab dimer. This highlighted in Figure 2D.

While the Fab and Fab'2 surely interact in the same manner independent of the Fc domain, they result in entirely different complexes at the same concentrations. From this result, we hypothesized that the interaction was a face-to-face blocking interaction involving the variable domains or HC/LC CDR network, which in the case of the monovalent Fab, would block any subsequent interactions. In the bivalent Fab'2 or mAb forms, however, one face-to-face interaction involving two molecules would still leave two available Fab domains for self-interaction and further self-oligomerization, and this holds true as the polymerization continues to arise in larger complexes in a concentration-dependent manner. In Figure 3, we highlight these proposed models that describe the self-interaction of the Fab domains that simply result in a dimer (Figure 3A), as well as the Fab'2 or mAb constructs that results in a growing oligomeric complex (Figure 3B). In the latter case, such a bivalent arrangement has the potential to dramatically affect rheological properties, and indeed in the case for the F104 mAb variant, incredibly large complexes are formed at merely 10 mg/mL in PBS pH 7.4 and high viscosities are achieved at or below 100 mg/mL (Figure 1).

A

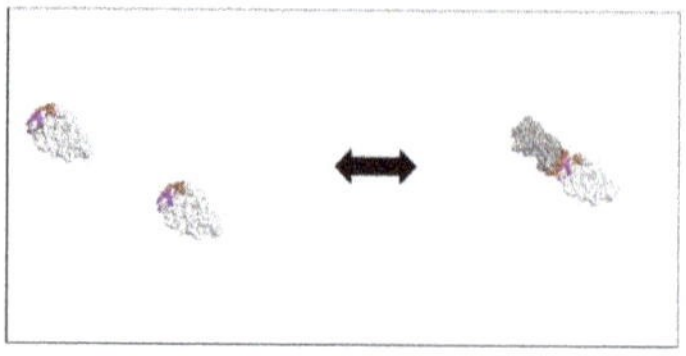

B

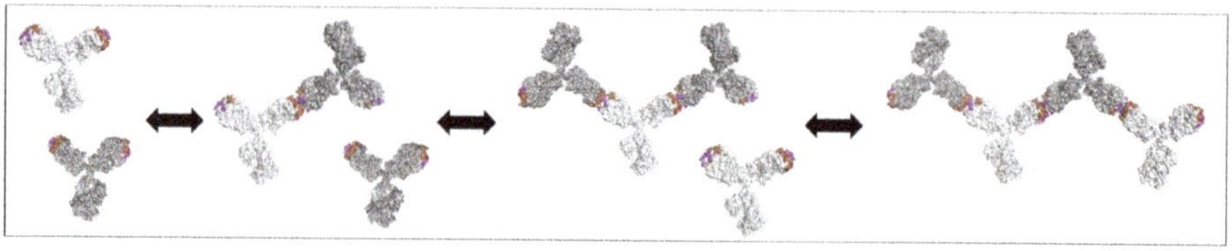

**Figure 3.** Model depicting (**A**) face-to-face interaction of the Fab forms yielding a dimer/monomer equilibrium and (**B**) bivalent mAb form yielding oligomerization and larger structures that alter bulk solution rheology (up to a tetramer shown).

### 3.3. *Evaluation of Self-Binding by BIAcore and ITC*

To evaluate the nature of this strong self-interaction from a binding perspective, antibody self-binding measurements were obtained using both a BIAcore (SPR) monovalent affinity assay and solution ITC (See Figure 4). In Figure 4A, the BIAcore experiment is diagrammed. Here, both the mAb and Fab were utilized; the mAb served as the ligand conjugated to a chip-bound anti-Fc antibody while the Fab was flowed as the analyte. Therefore, monovalent interactions between the Fab analyte and the mAb ligand could be measured. This was performed for both the HC-W104 and the strongly self-associating HC-F104 variant. By BIAcore SPR, no affinity was obtainable for the weak HC-W104 self-interaction, while a KD of 28 μM was measured for the HC-F104 variant (Figure 4C). This result is consistent with obtained DLS data for the F104 Fab, which was already partially self-associated at 1 mg/mL or ~20 μM. Using ITC, diagrammed in Figure 4B, W104 and F104 Fab preparations in PBS were fast diluted and the change in heat transfer in kcal/mol was measured over time. For the W104 Fab, a very weak KD of 5200 μM (5.2 mM) was obtained; for the F104 Fab, a KD of 120 μM was obtained. Similar to the monovalent self-affinity obtained by BIAcore SPR, the self-binding measured by solution ITC was in the low μM range. In the strongly self-interacting F104 system, a low μM monovalent self-affinity yields a highly negative diffusion interaction parameter obtained by DLS (−44.6 mL/g) and high viscosity at 100 mg/mL (43.7 cP).

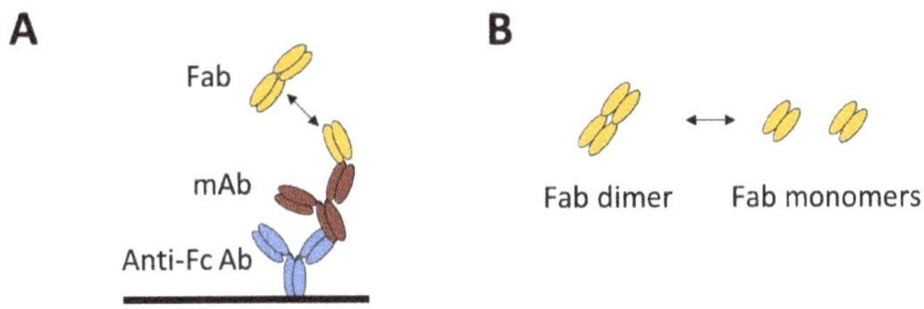


C

| Variant | BIAcore (SPR) | | | ITC |
|---|---|---|---|---|
| | ka (1/Ms) | kd (1/s) | KD (µM) | KD (µM) |
| **HC-W104** | NB | NB | NB | 5200 |
| **HC-F104** | $8.5 \times 10^4$ | 2.4 | 28 | 120 |

**Figure 4.** Schematics diagramming (**A**) BIAcore Surface Plasmon Resonance (SPR) format where the mAb serves as the ligand and the Fab of interest, shown here as an equilibrium of monomer and dimer forms, as the analyte. (**B**) Isothermal Calorimetry (ITC) format (in solution) where the Fab of interest is shown as an equilibrium of Fab dimer and monomer. In the experiment, the Fab is dimerized and is fast diluted to initially favor the monomer. (**C**) Measured affinities or KDs (nM) are tabulated for both approaches, including kd and ka for SPR.

### 3.4. Homology Modeling by Patch Analysis and Self-Docking

We performed structure-based modeling studies to explore potentially predictive and corrective tools, as well as in an effort to understand the dramatic difference in self-association observed between the F104 and W104 antibodies. Homology models of the F104 and W104 Fabs were built and subsequently used for computing hydrophobic and charged patches. Attraction between complementary patches, such as two hydrophobic patches or two oppositely charged ionic patches, can drive self-interaction and aggregation [29].

Analysis of the F104 Fab model identified a large hydrophobic patch with a surface area of 250 Å$^2$ that includes HC-F104 itself and eight additional residues in the LC and HC CDR regions (see Figure 5). This patch includes residues from multiple CDR loops in both the HC and LC, including HC-CDR3, LC-CDR1, and LC-CDR3. In evaluating this prominent hydrophobic patch, the top residues in order of contribution were HC-F104, LC-F92, LC-Y30, and LC-Y32.

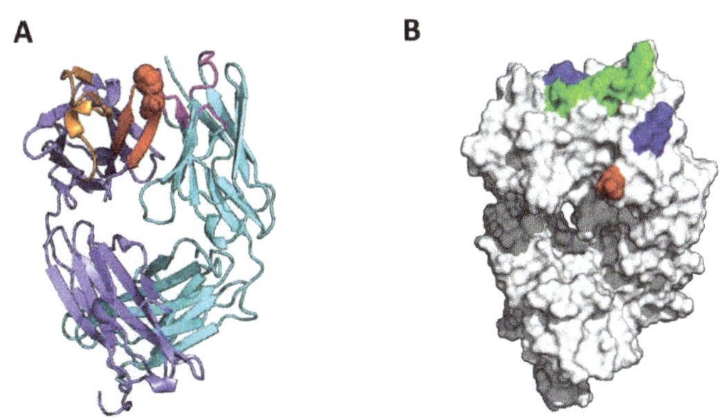


**Figure 5.** *Cont.*

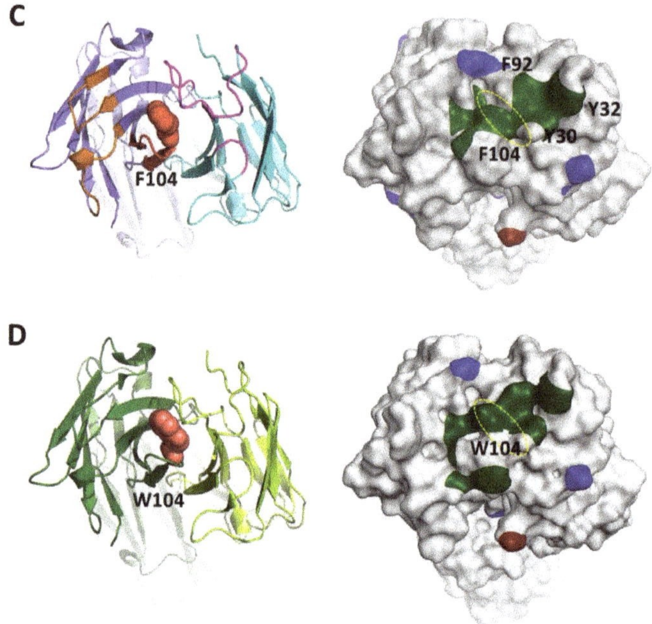


**Figure 5.** Fab homology models. (**A**) Model of HC:F104 Fab with VL in teal, VH in blue, VL CDR1-3 in purple, VH CDRs 1–2 in orange, VH CDR3 in red, and F104 in red spacefill. (**B**) Surface patch analysis of HC-F104 Fab homology model with hydrophobic patches in green, positive patches in blue, and negative patches in red. (**C**,**D**) Top view of CDR region of HC-F104 Fab and HC-W104 Fab homology models, respectively, rendered in Pymol. Left, ribbon depiction with F104 or W104 highlighted in red spacefill. Right, patch analysis rendered in MOE. F104 and W104 are circled with a dashed yellow ellipse.

Analysis of the W104 Fab homology model identified two hydrophobic patches in the CDR region with areas of 230 Å$^2$ and 60 Å$^2$. While the total hydrophobic patch size in the W104 antibody CDR region is larger than that of the F104 antibody, at 290 Å$^2$ versus 250 Å$^2$, the larger side chain of W104 and the presence of a polar NH moiety on the W104 side chain results in a disruption of the hydrophobic patch, which alone may contribute to a reduction in self-interaction in addition to changes in self-complementarity. Further, the larger W104 patch qualitatively differs from the single F104 patch, appearing more branched and discontinuous. In evaluating these patches in the W104 model, HC-W104, LC-F92, LC-Y30, and LC-Y32 are also found to be the top contributors to the overall hydrophobicity.

To generate hypotheses on specific molecular interactions contributing to the observed self-interaction, we performed protein–protein docking using two modeled F104 Fabs. Docking identified four plausible models for the interaction, which are shown in Figure 6. All models present a face-to-face interaction involving the HC- and LC-CDR apparatus. We prioritized model #3 because it has the largest interaction interface between the two Fabs. The model involves a nearly symmetric interface, with LC-Y30 and LC-F92 of one Fab forming stacking and hydrophobic interactions with HC-F104 of the opposing Fab. We note that docking with the W104 Fab model resulted in similar results, and so docking by itself is unable to rank order the two mutants.

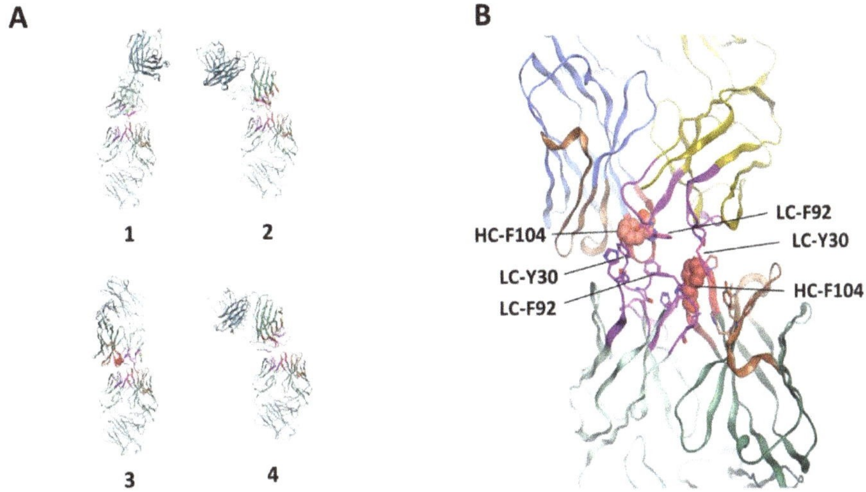


**Figure 6.** Fab protein-protein docking results. (**A**) Four poses seen in the top epitope cluster. (**B**) Details of model #3, a model for self-interaction supported by experimental data.

*3.5. Mutagenesis of Residues Revealed by Homology Model Patch Analysis and Self-Docking*

Position HC-104 clearly has a large impact on self-interaction and viscosity (Figures 1 and 2). In addition to this position, the preferred homology model revealed two additional residues involved in a prominent hydrophobic patch, LC-Y30 and LC-F92. A fourth residue, LC-Y32, was also identified but not was further evaluated in this study. To verify that these residues do indeed affect self-association in solution, various single mutants at positions LC-Y30 and LC-F92 were engineered on the strongly self-associating HC-F104 variant followed by expression and purification. Additional single point HC-104 mutants were also evaluated. Using purified variants, preparations of 1–10 mg/mL in PBS were evaluated by DLS. In Figure 7A, kD values obtained by DLS for all variants are plotted. Clearly, mutations at all sites had a significant impact on self-association, and some seemingly disrupted self-interaction to baseline kD values, similar to weakly associating W104 Fabs (−4.5 mL/g) and mAbs (−15.3 mL/g), as summarized in Figure 2A. These variants include LC-Y30D (−7.9 mL/g), LC-F92R (−10.8 mL/g), and HC-W104K (−4.8 mL/g), all of whose charged-based mutations were expected to disrupt the self-binding interface and significantly reduce the kD magnitude from the original value of −44.6 mL/g. Other mutations at position HC-104 to Arg and Asp also dramatically reduced the magnitude of kD values. These variants with low and acceptable kD's, plotted in Figure 7A, are shaded green. Interestingly, mutagenesis to Gly at all three sites resulted in largely negative kD values and only are slightly reduced in magnitude compared to the HC-F104 variant. Other mutations had a moderately reducing effect on self-association (e.g., HC-F104S and LC-F92H, shaded in black). Clearly, the self-interaction of this antibody was dramatically attenuated by single point mutations engineered across all three sites in both the HC- and LC-CDRs.

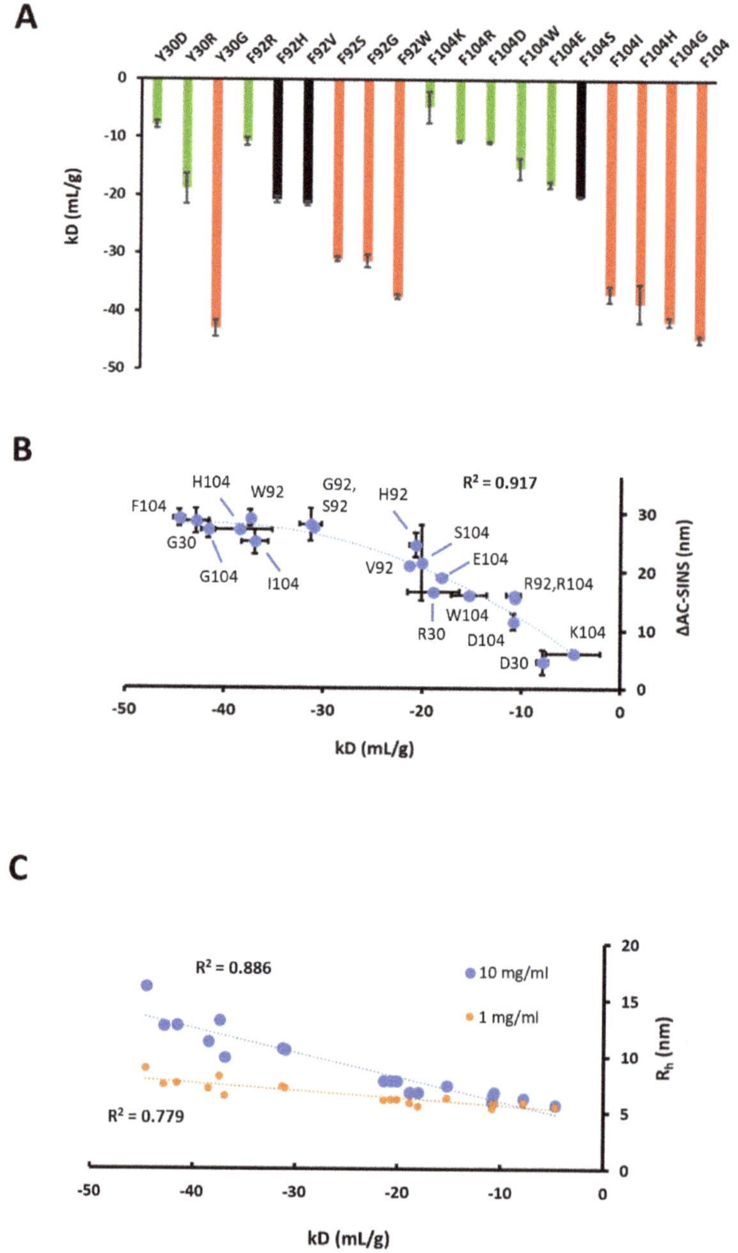


**Figure 7.** (**A**) kD (mL/g) obtained by DLS (1–10 mg/mL in PBS) shown as bar plots for all variants engineered at HC-104, LC-Y30, and LC-F92. Obtained kD values less than −30 mL/g are shaded red bars, and those higher than −20 mL/g are shaded green. Values between −20 and −30 mL/g are shaded black. Error bars are shown. (**B**) kD values obtained for all 19 interface mutants are plotted against corresponding ΔAC-SINS values (blue circles with error bars for each method and the $R^2$ of the parabolic fit). Each data point is labeled by variant. (**C**) kD values obtained for all 19 interface mutants are plotted against corresponding $R_h$ values (nm) obtained at both 1 mg/mL (orange circles) and 10 mg/mL (blue circles) along with $R^2$ of the linear fits.

All variants were also evaluated by AC-SINS, a nanoparticle-based screening assay that utilizes only microgram quantities of material per sample. In PBS, AC-SINS delta wavelength red shift values relative to buffer-only controls were obtained and plotted against the corresponding kD values obtained for each variant. In Figure 7B, obtained kD values and AC-SINS strongly correlate, and linearly correlate, for variants of low or moderate self-association properties (at ranges of approximately −5 to −20 mL/g for kD or 5–20 nm for AC-SINS). For variants that strongly self-associate, AC-SINS correctly predicts this outcome for all variants but is less sensitive to differences in this group, as kD obtained by DLS seemingly is, therefore, flattening the correlation at highly negative kD values (−30 mL/g or less). Therefore, a parabolic correlation is observed between kD and AC-SINS across a very large range of kD values (−4.8 to −44.6 mL/g) obtained in this study for 19 total variants. In Figure 7C, these obtained kD values all linearly correlate with obtained $R_h$ values at both 1 and 10 mg/mL, suggesting that self-association can be screened or evaluated by DLS using a single-point concentration measurement as opposed to generating several datapoints to obtain kD.

Although the mutagenesis of key residues was primarily done to evaluate their true effect on self-interaction, from a developability perspective, engineering as a means to correct for undesirable self-association should also maintain desirable target binding affinity at a minimum. In Table 1, the monovalent affinities acquired by SPR against the target antigen are reported. The variant with high and unacceptable self-association, HC-F104, had a sub-nanomolar affinity of 0.35 nM. While several charge-based mutations were made primarily to break self-interaction, such as LC-Y30R, LC-F92R, and HC-F104D, other mutations were more conservative, such as LC-F92W. In this panel, where over 20 single-point mutations were evaluated, the vast majority either yielded non-binding variants or those with markedly reduced SPR monovalent binding to the target antigen. Changes or reductions in KD (nM) were generally driven by reductions in the dissociation rate (kd). The only mutation where binding is strictly maintained or improved was HC-F104W, where SPR binding is slightly improved from 0.35 to 0.16 nM. This mutation also dramatically reduced self-association (kD was reduced from −44.6 to −15.3 mL/g). Other mutations bound with decreased target binding, such as LC-Y30R (12.3-fold decrease), LC-Y30H (15.6-fold decrease), and LC-F92W (15.6-fold decrease), yet also remained in the low nM range. However, if such mutations also dramatically reduced self-association and low nM range target binding affinity was desirable or acceptable, such variants would be reasonable developability candidates. In the case of LC-Y30R, self-association was dramatically reduced to an acceptable level (kD = −18.9 mL/g) and low nM affinity (4.36 nM) was achieved. A similar variant, LC-Y30H, also achieved low nM affinity target binding, but with higher self-association ($R_h$ of 11.2 nm at 4 mg/mL versus 6.5 nm at 4 mg/mL for LC-Y30R), meaning it would not likely be as developable, but insufficient material was purified to obtain a full kD plot. A full list of average $R_h$ values obtained by DLS for all variants is available in Table S2.

**Table 1.** BIAcore SPR and DLS results summary table for all variants. Tabulated are SPR KD (nM) values, the ratio of the measured KD and the HC-F104 reference KD ($KD/KD_{ref}$), and kD (mL/g) values obtained by DLS. The reference F104 is highlighted gray. All LC-Y30 and LC-F92 variants are site-directed mutants of the F104 variant. The single F104W variant with comparable BIAcore SPR KD to the reference is highlighted green. Variants with 10–20× differences in binding relative to the variant are highlighted yellow, and those with greater than 20-fold difference in SPR binding or non-binding (NB) are highlighted red. kD (mL/g) measured by DLS is color coded in terms of degree of self-association (greater than −20 mL/g, green; −20 to −29 mL/g, yellow; −30 mL/g or less, red). "NA" denotes no measurement.

| Variant | KD (nM) | $KD/KD_{ref}$ | kD (mL/g) |
|---|---|---|---|
| F104 | 0.35 | 1 | −44.6 |
| F104W | 0.16 | 0.44 | −15.3 |
| F104I | 108 | 305 | −36.9 |
| F104H | 1372 | 3879 | −38.5 |
| F104D | NB | NB | −10.9 |
| F104K | NB | NB | −4.8 |
| F104E | NB | NB | −18.1 |
| F104S | NB | NB | −20.1 |
| F104G | NB | NB | −41.6 |
| F104R | NB | NB | −10.7 |
| Y30R | 4.36 | 12.3 | −18.9 |
| Y30H | 5.5 | 15.6 | NA |
| Y30N | 45.5 | 129 | NA |
| Y30D | 60.9 | 172 | −7.9 |
| Y30Q | 82.6 | 234 | NA |
| Y30G | 170 | 481 | −42.9 |
| F92W | 5.51 | 15.6 | −37.4 |
| F92H | 35.1 | 99.2 | −20.7 |
| F92V | 38.7 | 109 | −21.4 |
| F92R | 50.2 | 142 | −10.8 |
| F92S | 62.7 | 177 | −31 |
| F92G | 142 | 402 | −31.3 |

## 4. Discussion

Here, strong self-interaction is observed for HC-F104 mAb constructs, resulting in significantly high viscosities at 100 mg/mL in both low formulation pH and physiological pH conditions and highly negative self-interaction parameters by DLS (see Figures 1 and 2). While formulation dependence is important to self-association propensity, and why, in the case of the HC-F104 mAb, viscosity remains high and even increases at a lower pH and ionic strength (versus PBS pH 7.4) is of interest, this was not explored further in this study. Since high viscosity was found to persist in these formulations that differ by ~2 pH units and have different ionic strengths, we hypothesized that the self-interaction was mainly hydrophobically driven. Additionally, since self-association was robust in PBS, a buffer that is physiologically and process-relevant, is likely to better conformationally stabilize the molecule, and accommodates all experimental approaches herein; all further experimental work was performed in PBS buffer to be consistent across variants and analytical methods.

In Figure 2, it is shown by DLS measurements that individual F104 Fab and Fab'2 domains self-interact independent of the Fc domain. For the W104 variants, self-interaction is dramatically reduced, although a weaker self-interaction persists, and consistent with this observation, low viscosity up to 100 mg/mL across different buffer conditions is observed (Figure 1). For the F104 domain variants, self-association is dramatically enhanced in the bivalent mAb/Fab'2 forms versus the monovalent Fab form, since each self-interaction event leads to additional available CDR faces to propagate oligomerization (see Figure 3). On the other hand, F104 Fab self-interaction leads to dimerization, blocking all additional interactions. Since the HC-CDR3-104 position greatly modulated this interaction and subsequent oligomerization was blocked in the Fab form, a face-to-face interaction involving the CDR regions was modeled (see Figure 3). Interestingly, similar observations were previously made, where rheological differences between monovalent Fab and bivalent Fab'2/mAb have been observed [28,55], although herein we show specifically that the F104 Fab self-dimerizes and further oligomerization is blocked. In terms of the development of bispecifics or multi-specifics, this implies that screening self-association in the bivalent, monospecific forms prior to formatting may not translate at all to rheological effects in the multi-specific form, which is typically monovalent for each targeting arm. This also clearly showcases the challenges of developing even higher order monospecific, multivalent antibody therapeutics (such as tetravalent), which, if appreciably self-interact, will be even more susceptible to high viscosity and negative rheological outcomes directly arising from inherent multi-valency.

By evaluating binding affinities by both BIAcore SPR and ITC, we see that our self-interacting F104 antibody has low μM affinity (28 μM by a monovalent SPR assay and 120 μM by ITC), which appears to be largely driven by high dissociation rates (kd) in comparison to antigen binding (See Figure 4). Self-association of W104 variants are significantly weaker, and either low or non-binding is measured by ITC and SPR, respectively. For the F104 mAb, a low μM self-affinity is significant when considering that concentrations during process and formulation conditions may reach into the mM range, meaning that all recombinant antibodies will be self-associated and oligomerized into large, heterogenous structures. Therefore, we see exponentially high viscosity (43.7–58.3 cP) up to 100 mg/mL (or ~700 μM), precluding higher process and formulation concentrations and further rendering the molecule high risk and undevelopable. This also clearly explains the low viscosities observed for the W104 mAb (4.5–4.8 cP, Figure 1), which self-binds in the mM range, according to ITC. Previously, binding by ITC and BIAcore has been evaluated for a self-interacting antibody system [56], but no affinities were obtained. Only recently, self-binding affinities for strong self-interacting antibodies or antibody binding fragments have been measured using AUC and ITC methods [21]. Similar to the results obtained by Schrag and coworkers for Fab variants, our obtained binding affinities for Fab and/or mAb interactions were in the low μM range. However, in our study, we further correlate these low μM affinities to high viscosities under typical formulation conditions, which clearly render the molecules high risk or undevelopable. Interestingly, our obtained range of binding affinities of 28–120 μM for the F104 Fabs are comparable to other protein–protein interactions that are biologically relevant, such as signaling proteins involved in macromolecular complexation or homodimerization [57,58].

Patch analysis of a preferred homology model of the F104 Fab shows a prominent hydrophobic patch (250 Å$^2$, Figure 5), which is consistent with surface areas of other antibodies known to self-interact [16,21]. Within this hydrophobic patch, three hydrophobic, aromatic residues are present: HC-F104, LC-Y30 and LC-F92. We clearly showed that mutations to all 3 sites dramatically impacted self-association characteristics. Consistent with a predominantly hydrophobic self-binding interface, charge-based mutations drastically suppressed self-interaction. Therefore, patch analysis of the F104 Fab correctly predicted the hydrophobic patch primarily responsible for self-interaction and was helpful in identifying the residues most responsible for it. However, this same exercise reaches a similar conclusion when performed for the W104 Fab, which does not appreciably self-interact in solution. In other words, clearly the patch(es) defined by homology modeling can inform which residues to engineer to correct for potential self-association, but by itself cannot predict actual self-association. In addition to likely requiring a defined patch potentially amenable to self-binding, antibody self-complementarity is a clear requirement for self-binding that arises in negative rheological outcomes. This is well underscored by the strong propensity of F104 mAbs or Fab'2s to self-interact, whereas a highly conservative Phe to Trp mutation exhibits a completely different outcome. Perhaps the more discontinuous nature of the patches generated for W104 offers clues to the difference in self-association outcomes. More likely responsible for the great differences between the W104 and F104 variants is the presence of sensitive, tunable self-complementarity analogous to antibody–

antigen interactions, where even conservative single point mutations can affect target binding affinities orders of magnitude [59]. In our self-interacting system, we see the large differences between F104 and W104 by both SPR and ITC binding, as well as by DLS and AC-SINS, highlighting how highly specific self-interaction is affected by a simple F104 to W104 transition. Further, both these variants have sub-nM binding against the same biological target (Table 1), therefore they are likely to exist in similar conformations. Interestingly, the mutagenesis of a similar Trp in the HC-CDR3 of another antibody has been shown to dramatically affect self-association [32], highlighting the importance of aromatic CDR residues in self-association. Other aromatic and hydrophobic CDR residues have been reported to be involved in antibody self-association as well [21,55], underscoring that self-association is often driven by forces beyond electrostatics.

The docking of F104 Fabs (see Methods), followed by epitope clustering, resulted in the identification of one dominant epitope, represented by four binding modes, all involving a face-to-face CDR interaction involving the F104 of both Fabs (Figure 6A). Previously, docking has been studied in relation to antibody–antigen or ligand binding [60,61]. In our study, one particular dock model reveals an interaction involving the same patch shown to affect face-to-face self-interaction in solution (Figure 6B). We already mentioned that self-interaction is likely to proceed by a face-to-face arrangement; docking allows us to build atomic models of possible specific interactions consistent with this.

Fab–Fab self-interactions have been characterized in the context of evaluating crystal contacts of pre-existing structures of Fabs and Fab–antigen complexes, where interactions were observed between the Fab CDR region and the Fab Framework (FW) [16,62]. In other reports, Fab–Fc interactions that are isotype dependent have been observed [56]. Additionally, Schrag and coworkers solved a crystal structure of a self-association prone Fab dimer and the dimer interface was mediated by symmetric CDR interactions composed of predominantly aromatic contacts [21]. Our face-to-face self-interaction is also driven by multiple CDR aromatic residues and is likely to be a common configuration and mechanism, since the CDR apparatus has evolved and is designed for binding. Moreover, the top ranked docking structure with the highest buried surface area is a highly symmetrical face-to-face complex involving the CDR apparatus. Consistent with our biochemical and modeling results, Fab self-association reported by Schrag and coworkers were attributed to the binding symmetry and surface flexibility of the CDR apparatus [21]. Interestingly, mutagenesis to glycine at all three CDR sites only slightly lowered self-association, indicating that, while the self-binding interface is highly specific to certain interfacial residues, it may be compensated through increased local main chain flexibility, enabled by the presence of glycine residues. Additionally, the inherent binding proclivity, along with inherent flexibility of the CDRs, means that the self-binding of the CDR apparatus to itself is inevitable and one of the likely possibilities for self-interacting recombinant antibodies.

From a developability perspective, there is significant interest in understanding how to better predict and correct self-association at a sequence or in silico level without impacting other key attributes such as binding and stability. Here, simply using a homology model, three sites were revealed in a prominent hydrophobic patch and all were shown to significantly impact self-interaction. To be viable correction variants, at a minimum target antigen, binding affinity or activity must be acceptably maintained. In Figure 7, only one variant out of the 22 total variants that were mutated at three separate CDR sites, HC-W104F, maintains both sub-nM target antigen affinity and significantly lowers self-interaction to an acceptable range. Another variant, LC-Y30R, also significantly lowered self-association to an acceptable level and maintained low nM target affinity. This exercise demonstrates the difficulties of correcting problematic self-association, particularly when the same CDR apparatus that has evolved for binding is also involved in strong self-interaction. Often, high affinity, or significantly improving it through affinity maturation, increases the likelihood of further encountering negative developability attributes [21,63]. Supporting this notion, in the case of the HC-F104 variant, we have both sub-nM target binding and unacceptable self-association propensity. Because antibody self-interaction will almost certainly involve the functional CDR apparatus, possibly on both molecules, the probability of success for each variant to maintain tight target binding and reduce self-association is low (<10% in our study). Therefore, we have learned that the selection and evaluation of more variants at various positions will increase the likelihood of achieving the desired results. Overall, it is shown here that both homology patch analysis and self-docking can inform potential self-associating regions with low μM affinity and residues amenable to corrective engineering despite its difficulty, but also may not predict actual self-association because of the apparent requirement for unique self-interaction specificity or complementarity between molecules.

## 5. Conclusions

Herein, reversible antibody self-association is evaluated biochemically using individual domains, through binding affinity measurements, by in silico homology modeling and docking approaches, and by site-directed mutagenesis. Interestingly, the self-interaction of mAb variants are shown to likely proceed through a blocking face-to-face interaction involving the HC and LC CDRs that is potently magnified by bivalency. This self-binding, leading to unacceptable viscosity at 100 mg/mL in different formulations, also has low μM self-affinity and is highly tunable by and sensitive to single-point mutations to three different HC/LC CDR residues informed by homology model patch analysis and self-docking. The evaluation of these self-association mutants demonstrate that corrective engineering is difficult yet feasible and warrants several mutations at multiple positions to increase likelihood of success. Additionally, in silico approaches, while useful and informative in identifying potential self-interacting regions, are likely not able to address the specificity and complementarity of self-binding that is evident in a system that exhibits significant self-interaction propensity and sensitivity to even conservative mutations at multiple sites. This study enhances our understanding of antibody self-association and potential means of addressing and mitigating it from a developability perspective.

**Supplementary Materials:** The following are available online at https://www.mdpi.com/2073-4468/10/1/8/s1, Table S1: SE-UPLC retention times (RT) and % purity tabulated for each variant, Table S2: Tabulated diffusional coefficients (Do), hydrodynamic radii (Rh) measured over the range of 1–10 mg/mL, and ΔAC-SINS values (in nm) for all variants evaluated. "NA" denotes not measured or available.

**Author Contributions:** Conceptualization, C.M., A.C.; methodology, C.M., A.C., M.H.; formal analysis, C.M., A.C., T.F., M.H., J.B., M.U., G.R.; investigation, C.M., A.C., T.F., M.H.; writing—original draft preparation, C.M.; writing—review and editing, C.M., A.C., L.F.-D.; supervision, C.S., L.F.-D. All authors have read and agreed to the published version of the manuscript.

**Funding:** This research received no external funding.

**Institutional Review Board Statement:** Not applicable.

**Informed Consent Statement:** Not applicable.

**Data Availability Statement:** Not applicable.

**Conflicts of Interest:** The authors, all employees of Merck & Co. Inc., declare no conflict of interest.

## References

1. Lu, R.; Hwang, Y.; Liu, I.; Lee, C.C.; Tsai, H.Z.; Li, H.J.; Wu, H.C. Development of therapeutic antibodies for the treatment of diseases. *J. Biomed. Sci.* **2020**, *27*, 1–30. [CrossRef]
2. Kaplon, H.; Reichert, J.M. Antibodies to watch in 2019. *mAbs* **2019**, *11*, 219–238. [CrossRef]
3. Shire, S.J. Formulation and manufacturability of biologics. *Curr. Opin. Biotechnol.* **2009**, *20*, 708–714. [CrossRef] [PubMed]
4. Le Basle, Y.; Chennell, P.; Tokhadze, N.; Astier, A.; Sautou, V. Physicochemical stability of monoclonal antibodies: A review. *J. Pharm. Sci.* **2020**, *109*, 169–190. [CrossRef] [PubMed]
5. Jain, T.; Sun, T.; Durand, S.; Hall, A.; Houston, N.R.; Nett, J.H.; Sharkey, B.; Bobrowitcz, B.; Caffry, I.; Yu, Y.; et al. Biophysical features of the clinical mAb landscape. *Proc. Natl. Acad. Sci. USA* **2017**, *114*, 944–949. [CrossRef]
6. Bailly, M.; Mieczkowski, C.; Juan, V.; Metwally, E.; Tomazela, D.; Baker, J.; Uchida, M.; Kofman, E.; Raoufi, F.; Motlagh, S.; et al. Predicting antibody developability profiles through early stage discovery screening. *mAbs* **2020**, *12*, 1–28. [CrossRef] [PubMed]

7. Wang, W.; Singh, S.; Zeng, D.; King, K.; Nema, S. Antibody structure, instability, and formulation. *J. Pharm. Sci.* **2007**, *96*, 1–26. [CrossRef]
8. Jarasch, A.; Koll, H.; Regula, J.T.; Bader, M.; Papadimitriou, A.; Kettenberger, H. Developability assessment during the selection of novel therapeutic antibodies. *J. Pharm. Sci.* **2015**, *104*, 1885–1898. [CrossRef]
9. Daugherty, A.L.; Mrsny, R.J. Formulation and delivery issues for monoclonal antibody therapeutics. *Adv. Drug Deliv. Rev.* **2006**, *58*, 686–706. [CrossRef]
10. Avery, L.B.; Wade, J.; Wang, M.; Tam, A.; King, A.; Piche-Nicholas, N.; Kavosi, M.S.; Penn, S.; Cirelli, D.; Kurz, J.C.; et al. Establishing in vitro in vivo correlations to screen monoclonal antibodies for physicochemical properties related to favorable human pharmacokinetics. *mAbs* **2018**, *10*, 244–255. [CrossRef]
11. Hotzel, I.; Theil, F.P.; Bernstein, L.J.; Prabhu, S.; Deng, R.; Quintana, L.; Lutman, J.; Sibia, R.; Chan, P.; Bumbaca, D.; et al. A strategy for risk mitigation of antibodies with fast clearance. *mAbs* **2012**, *4*, 753–760. [CrossRef]
12. Kelly, R.; Sun, T.; Jain, T.; Caffry, I.; Yu, Y.; Cao, Y.; Lynaugh, H.; Brown, M.; Vásquez, M.; Wittrup, K.D.; et al. High throughput cross-interaction measures for human IgG1 antibodies correlate with clearance rates in mice. *mAbs* **2015**, *7*, 770–777. [CrossRef]
13. Geng, S.B.; Cheung, J.K.; Narasimhan, C.; Shameen, M.; Tessier, P.M. Improving monoclonal antibody selection and engineering using measurements of colloidal protein interactions. *J. Pharm. Sci.* **2014**, *103*, 3356–3363. [CrossRef]
14. Liu, J.; Nguyen, M.D.; Andya, J.D.; Shire, S.J. Reversible self-association increases the viscosity of a concentrated monoclonal antibody in aqueous solution. *J. Pharm. Sci.* **2005**, *94*, 1928–1940. [CrossRef]
15. Laue, T.M.; Shire, S.J. The molecular interaction process. *J. Pharm. Sci.* **2020**, *109*, 154–160. [CrossRef] [PubMed]
16. Bethea, D.; Wu, S.J.; Luo, J.; Hyun, L.; Lacy, E.R.; Teplyakov, A.; Jacobs, S.A.; O'Neil, K.T.; Gilliland, G.L.; Feng, Y. Mechanisms of self association of a human monoclonal antibody CNTO607. *Protein Eng. Des. Sel.* **2012**, *25*, 531–537. [CrossRef] [PubMed]
17. Saluja, A.; Kalonia, D.S. Nature and consequences of protein–protein interactions in high protein concentration solutions. *Int. J. Pharm.* **2008**, *358*, 1–15. [CrossRef]
18. Tomar, D.S.; Kumar, S.; Singh, S.K.; Goswami, S.; Li, L. Molecular basis of high viscosity in concentrated antibody solutions: Strategies for high concentration drug product development. *mAbs* **2016**, *8*, 216–218. [CrossRef] [PubMed]
19. Cromwell, M.; Felten, C.; Flores, H.; Liu, J.; Shire, S.J. Self-association of therapeutic proteins: Implications for product development. In *Misbehaving Proteins*; Murphy, R.M., Tsai, A.M., Eds.; Springer: Berlin/Heidelberg, Germany, 2006; pp. 313–330.
20. Warne, N.W. Development of high concentration protein biopharmaceuticals: The use of platform approaches in formulation development. *Eur. J. Pharm. Biopharm.* **2009**, *78*, 208–212. [CrossRef] [PubMed]
21. Schrag, J.D.; Picard, M.E.; Gaudreault, F.; Gagnon, L.P.; Baardsnes, J. Binding symmetry and surface flexibility mediate antibody self-association. *mAbs* **2019**, *11*, 1300–1318. [CrossRef] [PubMed]
22. Shire, S.J.; Shahrokh, Z.; Liu, J. Challenges in the development of high protein concentration formulations. *J. Pharm. Sci.* **2004**, *93*, 1390–1402. [CrossRef]
23. Lehermayr, C.; Mahler, H.C.; Mader, K.; Fischer, S. Assessment of net charge and protein-protein interactions of different monoclonal antibodies. *J. Pharm. Sci.* **2011**, *100*, 2551–2562. [CrossRef] [PubMed]
24. Yadav, S.; Laue, T.M.; Kalonia, D.S.; Singh, S.N.; Shire, S.J. The influence of charge distribution on self-association and viscosity behavior of monoclonal antibody solutions. *Mol. Pharm.* **2012**, *9*, 791–802. [CrossRef] [PubMed]
25. Esfandiary, R.; Parupudi, A.; Casas-Finet, J.; Gadre, D.; Sathish, H. Mechanism of reversible self-association of a monoclonal antibody: Role of electrostatic and hydrophobic interactions. *J. Pharm. Sci.* **2014**, *104*, 577–586. [CrossRef]
26. Salinas, B.A.; Sathish, H.A.; Bishop, S.M.; Harn, N.; Carpenter, J.F.; Randolph, T.W. Understanding and modulating opalescence and viscosity in a monoclonal antibody formulation. *J. Pharm. Sci.* **2010**, *99*, 82–93. [CrossRef] [PubMed]
27. Casaz, P.; Boucher, E.; Wollacott, R.; Pierce, B.G.; Rivera, R.; Sedic, M.; Ozturk, S.; Thomas, W.D., Jr.; Wang, Y. Resolving self-association of a therapeutic antibody by formulation optimization and molecular approaches. *mAbs* **2014**, *6*, 1533–1539. [CrossRef]
28. Kanai, S.; Liu, J.; Patapoff, T.W.; Shire, S.J. Reversible self-association of a concentrated monoclonal antibody solution mediated by Fab-Fab interaction that impacts solution viscosity. *J. Pharm. Sci.* **2008**, *97*, 4219–4227. [CrossRef] [PubMed]
29. Agrawal, N.; Helk, B.; Kumar, S.; Mody, N.; Sathish, H.A.; Samra, H.S.; Buck, P.M.; Li, L.; Trout, B.L. Computational tool for the early screening of monoclonal antibodies for their viscosities. *mAbs* **2016**, *8*, 43–48. [CrossRef] [PubMed]
30. Apgar, J.R.; Tam, A.S.P.; Sorm, R.; Moesta, S.; King, A.C.; Yang, H.; Kelleher, K.; Murphy, D.; D'Antona, A.M.; Yan, G.; et al. Modeling and mitigation of high concentration antibody viscosity through structure-based computer-aided protein design. *PLoS ONE* **2020**, *5*, 1–26. [CrossRef] [PubMed]
31. Li, L.; Kumar, S.; Buck, P.M.; Burns, C.; Lavoie, J.; Singh, S.K.; Warne, N.W.; Nichols, P.; Luksha, N.; Boardman, D. Concentration dependent viscosity of monoclonal antibody solutions: Explaining experimental behavior in terms of molecular properties. *Pharm. Res.* **2014**, *31*, 3161–3178. [CrossRef]
32. Shan, L.; Mody, N.; Sormani, P.; Rosenthal, K.L.; Damschroder, M.M.; Esfandiary, R. Developability assessment of engineered monoclonal antibody variants with a complex self-association behavior using complementary analytical and in silico tools. *Mol. Pharm.* **2018**, *15*, 5697–5710. [CrossRef]
33. Raybould, M.I.J.; Marks, C.; Krawczyk, K.; Taddese, B.; Nowak, J.; Lewis, A.P.; Bujotzek, A.; Shi, J.; Deane, C.M. Five computational developability guidelines for therapeutic antibody profiling. *Proc. Natl. Acad. Sci. USA* **2019**, *116*, 4025–4030. [CrossRef]

34. Sharma, V.K.; Patapoff, T.W.; Kabakoff, B.; Pai, S.; Hilario, E.; Zhang, B.; Li, C.; Borisov, O.; Kelley, R.F.; Chorny, I.; et al. In silico selection of therapeutic antibodies for development: Viscosity, clearance, and chemical stability. *Proc. Natl. Acad. Sci. USA* **2014**, *111*, 18601–18606. [CrossRef]
35. Cromwell, M.E.; Hilario, E.; Jacobson, F. Protein aggregation and bioprocessing. *AAPS J.* **2006**, *8*, E572–E579. [CrossRef]
36. Monkos, K. Concentration and temperature dependence of viscosity in lysozyme aqueous solutions. *Biochim. Biophys. Acta* **1997**, *1339*, 304–310. [CrossRef]
37. Yadav, S.; Shire, S.J.; Kalonia, D.S. Factors affecting the viscosity in high concentration solutions of different monoclonal antibodies. *J. Pharm. Sci.* **2010**, *99*, 4812–4826. [CrossRef]
38. Scherer, T.M.; Liu, J.; Shire, S.J.; Minton, A.P. Intermolecular interactions of IgG1 monoclonal antibodies at high concentrations characterized by light scattering. *J. Phys. Chem. B* **2010**, *114*, 12948–12957. [CrossRef] [PubMed]
39. Saito, S.; Hasegawa, J.; Kobayashi, N.; Kishi, N.; Uchiyama, S.; Fukui, K. Behavior of monoclonal antibodies: Relation between the second virial coefficient (B(2)) at low concentrations and aggregation propensity and viscosity at high concentrations. *Pharm. Res.* **2012**, *29*, 397–410. [CrossRef] [PubMed]
40. Borwankar, A.U.; Dear, B.J.; Twu, A.; Hung, J.J.; Dinin, A.K.; Wilson, B.K.; Yue, J.; Maynard, J.A.; Truskett, T.M.; Johnston, K.P. Viscosity reduction of a concentrated monoclonal antibody with arginine·HCl and arginine·glutamate. *Ind. Eng. Chem. Res.* **2016**, *55*, 11225–11234. [CrossRef]
41. Inoue, N.; Takai, E.; Arakawa, T.; Shiraki, K. Specific decrease in solution viscosity of antibodies by arginine for therapeutic formulations. *Mol. Pharm.* **2014**, *11*, 1889–1896. [CrossRef]
42. Connolly, B.D.; Petry, C.; Yadav, S.; Demeule, B.; Ciaccio, N.; Moore, J.M.; Shire, S.J.; Gokarn, Y.R. Weak interactions govern the viscosity of concentrated antibody solutions: High-throughput analysis using the diffusion interaction parameter. *Biophys. J.* **2012**, *103*, 69–78. [CrossRef] [PubMed]
43. Saluja, A.; Fesinmeyer, R.M.; Hogan, S.; Brems, D.N.; Gokarn, Y.R. Diffusion and sedimentation interaction parameters for measuring the second virial coefficient and their utility as predictors of protein aggregation. *Biophys. J.* **2010**, *99*, 2657–2665. [CrossRef] [PubMed]
44. Tessier, P.M.; Wu, J.; Dickinson, C.D. Emerging methods for identifying monoclonal antibodies with low propensity to self-associate during the early discovery process. *Expert Opin. Drug Deliv.* **2014**, *11*, 461–465. [CrossRef]
45. Geng, S.B.; Wittekind, M.; Vigil, A.; Tessier, P.M. Measurements of monoclonal antibody self-association are correlated with complex biophysical properties. *Mol. Pharm.* **2016**, *13*, 1636–1645. [CrossRef]
46. Sule, S.V.; Dickinson, C.D.; Lu, J.; Chow, C.K.; Tessier, P.M. Rapid analysis of antibody self-association in complex mixtures using immunogold conjugates. *Mol. Pharm.* **2013**, *10*, 1322–1331. [CrossRef]
47. Jayaraman, J.; Wu, J.; Brunelle, M.C.; Cruz, A.M.M.; Goldberg, D.S.; Lobo, B.; Shah, A.; Tessier, P.M. Plasmonic measurements of monoclonal antibody self-association using self-interaction nanoparticle spectroscopy. *Biotechnol. Bioeng.* **2014**, *111*, 1513–1520. [CrossRef]
48. Howlett, G.J.; Minton, A.P.; Rivas, G. Analytical ultracentrifugation for the study of protein association and assembly. *Curr. Opin. Chem. Biol.* **2006**, *10*, 430–436. [CrossRef]
49. Jimenez, M.; Rivas, G.; Minton, A.P. Quantitative characterization of weak self-association in concentrated solutions of immunoglobulin G via the measurement of sedimentation equilibrium and osmotic pressure. *Biochemistry* **2007**, *46*, 8373–8378. [CrossRef]
50. Esfandiary, R.; Hayes, D.B.; Parupudi, A.; Casas-Finet, J.; Bai, S.; Samra, H.S.; Shah, A.U.; Sathishi, H.A. A Systematic multitechnique approach for detection and characterization of reversible self-association during formulation development of therapeutic antibodies. *J. Pharm. Sci.* **2012**, *102*, 62–72. [CrossRef] [PubMed]
51. Liu, Y.; Caffry, I.; Wu, J.; Geng, S.B.; Jain, T.; Sun, T.; Reid, F.; Cao, Y.; Estep, P. High-throughput screening for developability during early-stage antibody discovery using self-interaction nanoparticle spectroscopy. *mAbs* **2014**, *6*, 483–492. [CrossRef] [PubMed]
52. Wu, S.J.; Luo, J.; O'Neil, K.T.; Kang, J.; Lacy, E.R.; Canziani, G.; Baker, A.; Huang, M.; Tang, Q.M.; Raju, T.S.; et al. Structure-based engineering of a monoclonal antibody for improved solubility. *Protein Eng. Des. Sel.* **2010**, *23*, 643–651. [CrossRef]
53. Brown, W.; Nicolai, T. *Dynamic Light Scattering: The Method and Some Applications*; Oxford University Press: New York, NY, USA, 1993.
54. Berteau, C.; Filipe-Santos, O.; Wang, T.; Rojas, H.E.; Granger, C.; Schwarzenbach, F. Evaluation of the impact of viscosity, injection volume, and injection flow rate on subcutaneous injection tolerance. *Med. Devices Evid. Res.* **2015**, *8*, 473–484.
55. Tilegenova, C.; Izadi, S.; Yin, J.; Huang, C.S.; Wu, J.; Ellerman, D.; Hymowitz, S.G.; Walters, B.; Salisbury, C.; Carter, P.J. Dissecting the molecular basis of high viscosity of monospecific and bispecific IgG antibodies. *mAbs* **2020**, *12*, 1–11. [CrossRef]
56. Nishi, H.; Miyajima, M.; Wakiyama, N.; Kubota, K.; Hasegawa, J.; Uchiyama, S.; Fukui, K. Fc domain mediated self-association of an IgG1 monoclonal antibody under a low ionic strength condition. *J. Biosci. Bioeng.* **2011**, *112*, 326–332. [CrossRef] [PubMed]
57. Kuriyan, J.; Eisenberg, D. The origin of protein interactions and allostery in colocalization. *Nature* **2007**, *450*, 983–990. [CrossRef]
58. Kastritis, P.L.; Bonvin, A.M. On the binding affinity of macromolecular interactions: Daring to ask why proteins interact. *J. R. Soc. Interface* **2013**, *10*, 1–27. [CrossRef] [PubMed]
59. Chen, C.; Roberts, V.A.; Rittenberg, M.B. Random point mutations in an antibody CDR2 sequence: Many mutated antibodies lose their ability to bind antigen. *J. Exp. Med.* **1992**, *176*, 855–866. [CrossRef]

60. Ambrosetti, F.; Jimenez-Garcia, B.; Roel-Touris, J.; Bonvin, A.M. Modeling antibody-antigen complexes by information-driven docking. *Structure* **2020**, *28*, 119–129. [CrossRef] [PubMed]
61. Pedotti, M.; Simonelli, L.; Livoti, E.; Varani, L. Computational docking of antibody-antigen complexes, opportunities and pitfalls illustrated by influenza hemagglutinin. *Int. J. Mol. Sci.* **2011**, *12*, 226–251. [CrossRef]
62. Teplyakov, A.; Obmolova, G.; Wu, S.J.; Luo, J.; Kang, J.; O'Neil, K.; Gilliland, G.L. Epitope mapping of anti-interleukin-13 neutralizing antibody CNTO607. *J. Mol. Biol.* **2009**, *389*, 115–123. [CrossRef]
63. Shetata, L.; Maurer, D.P.; Wec, A.Z.; Lilov, A.; Champney, A.; Sun, T.; Archambault, K.; Burnina, I.; Lynaugh, H.; Zhi, X.; et al. Affinity maturation enhances antibody specificity but compromises conformational stability. *Cell Rep.* **2019**, *28*, 3300–3308. [CrossRef] [PubMed]

antibodies

MDPI

*Article*

# Selection and Characterization of Single-Domain Antibodies for Detection of Lassa Nucleoprotein

**George P. Anderson, Jinny L. Liu, Lisa C. Shriver-Lake and Ellen R. Goldman ***

Naval Research Laboratory, Center for Biomolecular Science and Engineering, Washington, DC 20375, USA; george.anderson@nrl.navy.mil (G.P.A.); jinny.liu@nrl.navy.mil (J.L.L.); lisa.shriverlake@nrl.navy.mil (L.C.S.-L.)
* Correspondence: ellen.goldman@nrl.navy.mil; Tel.: +1-202-404-6052

Received: 7 October 2020; Accepted: 7 December 2020; Published: 17 December 2020

**Abstract:** Lassa virus is the etiologic agent of Lassa fever, an acute and often fatal illness endemic to West Africa. It is important to develop new reagents applicable either for the specific diagnosis or as improved therapeutics for the treatment of Lassa fever. Here, we describe the development and initial testing of llama-derived single-domain antibodies that are specific for the Lassa virus nucleoprotein. Four sequence families based on complementarity-determining region (CDR) homology were identified by phage-based enzyme-linked immunosorbent assays, however, the highest affinity clones all belonged to the same sequence family which possess a second disulfide bond between Framework 2 and CDR3. The affinity and thermal stability were evaluated for each clone. A MagPlex-based homogeneous sandwich immunoassay for Lassa virus-like particles was also demonstrated to show their potential for further development as diagnostic reagents.

**Keywords:** single-domain antibody; Lassa virus; Arenaviridae family; MagPlex

## 1. Introduction

Lassa fever, a hemorrhagic fever caused by Lassa virus (LASV), is endemic in West Africa [1]. There are numerous febrile illnesses common in West Africa, which makes Lassa fever diagnosis based solely on clinical symptoms impossible. Thus, laboratory diagnostics play a critical role in stemming the spread of this serious disease and thereby hopefully quenching an epidemic at its earliest stage. Additionally, diagnosis of Lassa fever is critical for limiting nosocomial infections, especially in the maternity ward where prospective mothers and their unborn children are at an extremely high risk of death if infected [2,3].

Several factors have served to limit the impact of Lassa fever diagnostics. Many Lassa fever patients are asymptomatic or have nonspecific symptoms [1]. This necessitates the testing of nearly all patients, creating a strain on healthcare costs. This is compounded by the fact that developing a reliable assay is complicated by the extent of LASV sequence diversity [4]. An additional challenge is the need for the safe collection and handling of specimens to prevent infection of the medical staff. The lack of high-containment laboratories in the areas where the need is greatest has led to limited Lassa fever assay development and validation studies [5].

While viral culture has been the "gold standard" for the diagnosis of Lassa fever, nucleic acid detection methods have become the mainstay approach for many laboratory-based diagnostics. This type of assay can have unparalleled sensitivity and specificity. While costs of nucleic acid-based testing were initially high, automation and miniaturization have been bringing costs down each year. However, probe design is challenging due to the high sequence variability of the various strains of LASV [5].

Alternatively, antigen and antibody tests have been developed for LASV, which are attractive given the cost constraints and the austere conditions where these tests are most often required [6–8].

Both antigen and antibody tests have their strengths and weaknesses, as their ability to adequately identify virus is dependent on the number of days which have elapsed between infection and sample collection. Thus, for adequate identification, it will likely be necessary to develop multiplexed or parallel assays where a number of targets are assayed for simultaneously. Most antigen tests are based on the detection of LASV nucleoprotein (NP). NP antigen is detectable for the first week following infection with Lassa fever, but wanes during the second week as the levels of immunoglobulins increase [9]. Fatal cases of Lassa fever exhibited higher levels of NP versus nonfatal cases, suggesting utility for monitoring the course of the infection that one day might aid in selection of treatments [10]. Since NP is present earlier during the course of infection, it yields an earlier diagnosis than antibody detection, and represents the actual presence of virus, thus providing actionable information. Antibody, on the other hand, can be present in patients long after the infection has resolved, meaning detection of antibody is not always indicative of an active infection, just as absence of NP does not rule out a persistent viremia, since NP levels become undetectable over time [8].

Currently, the antibody-based detection of NP has utilized conventional antibodies [6,11]. While effective, these reagents cannot easily be genetically manipulated and tailored for optimization with specific detection platforms. Furthermore, polyclonal antibodies can show variability in activity from lot to lot, while monoclonal antibodies are produced by cell lines that require careful maintenance to prevent loss. We have sought to augment conventional antibodies with recombinant single-domain antibodies (sdAb) derived from the immune repertoire of an immunized llama. In addition to conventional antibodies, camelids, including llamas, possess classes of heavy-chain-only antibodies where the binding is mediated by an unpaired variable heavy domain (VHH) [12]. This simplifies the creation of recombinant binding molecules as the binding requires only the single variable domain, not the pairing of the variable heavy and light domains as in conventional antibodies. Recombinantly expressed VHH are termed sdAb or nanobodies [13].

These sdAb also possess other advantages, such as being more thermally stable, having the ability to refold following thermal denaturation, and being easier to recombinantly produce than most single-chain antibodies (recombinantly derived binding domains from conventional antibodies) [14]. Herein, we describe in detail the selection and evaluation of sdAb specific for the LASV NP, which could, with continued optimization of their functionality, be of value for the detection and diagnosis of Lassa fever.

## 2. Materials and Methods

### *2.1. Reagents*

Most reagents were obtained from VWR International (Radnor, PA, USA), Sigma Aldrich (St. Louis, MO, USA), or Thermo Fisher Scientific (Waltham, MA, USA). Cloning reagents were purchased from New England Biolabs (Ipswich, MA, USA). Zalgen (Germantown, MD, USA) was the source of the recombinant NPn (N-terminal portion of NP encoding residues 1-340), NPc (C-terminal portion of NP encoding residues 341–569), and Lassa virus-like particles (VLPs). All LASV reagents were based on LASV lineage IV (Josiah strain).

### *2.2. Selection and Production of sdAb*

A llama was immunized with Lassa NPn (N-terminal portion of NP) under contract with Triple J Farms, Bellingham, WA. Triple J Farms has an active Institutional Animal Care and Use Committee which reviews all immunization protocols, and approved this protocol. Anticoagulated blood was shipped overnight and buffy coat cells were isolated by centrifugation using Novamed Uni-Sep tubes. RNA was isolated from which cDNA was produced, amplifying the mRNA of the variable heavy domains, from which a phage display library was constructed as described previously [15]. Panning on NPn protein adsorbed in the wells of a microtiter plate was performed as described [15]. Using both monoclonal phage enzyme-linked immunosorbent assay (ELISA) and monoclonal phage MagPlex assay, potentially positive clones were identified after rounds two and three [16].

To facilitate periplasmic production, the coding sequences for the sdAb were cloned from the pecan21 [17] phage display vector into pET22b [18]. To produce the sdAb, expression plasmids were transformed into Tuner (DE3). Cultures (50 mL in terrific broth [TB] with 100 µg/mL ampicillin) were started from freshly transformed colonies and grown in shake-flasks overnight at 25 °C. The next day, each culture was added to 450 mL of TB with ampicillin and incubated with shaking for 2 h at 25 °C. Cultures were then induced using isopropyl-D-1 thiogalactoside (0.5 mM) and incubated for another 2 h.

Purification of sdAb was through an osmotic shock protocol followed by immobilized metal affinity chromatography (IMAC) and fast protein liquid chromatography as described previously [19]. Nearly all the protein eluted from the gel filtration column as single peak consistent with the sdAb being a monomer; very few contaminants remained following IMAC chromatography. UV absorption was utilized to determine sdAb concentration, and sdAb preparations were refrigerated for immediate use or frozen for long-term storage.

### 2.3. *Surface Plasmon Resonance*

A ProteOn XPR36 (Bio-Rad, Hercules, CA, USA) was used to measure the binding kinetics of the sdAb. Lanes of a GLC chip were immobilized with Lassa NPn and NPc from Zalgene diluted to 20 µg/mL in 10 mM sodium acetate pH 5.0, as described previously [20]. The binding kinetics of each sdAb were determined by flowing a range of concentrations of each sdAb over the chip to produce an array of binding curves. The chip was regenerated using 0.085% phosphoric acid between data sets. Data were corrected for interspot and zero concentration. The ProteOn Manager 2.1 software was employed for analysis of the data using the Langmuir model. The standard error of the fits was less than 10%.

### 2.4. *Determining Melting Temperature and Refolding by Circular Dichroism*

Circular dichroism (CD) was performed using a Jasco J-815 Spectropolarimeter, following a protocol used in prior work [15]. The sdAb samples were diluted to 15 µg/mL in deionized water and placed in a quartz cuvette with 1 cm path length and CD was measured at an ultraviolet wavelength between 204 and 206 nm. The sdAb samples were heated from 25 °C to 95 °C at a rate of 2.5 °C/min and then cooled back to 25 °C at the same rate.

### 2.5. *MagPlex Direct Binding and Homogeneous Sandwich Assays*

Specificity and an indication of affinity were appraised via the direct binding of the sdAb to Lassa NPn recombinant protein immobilized on MagPlex magnetic microspheres (Luminex, Austin, TX, USA). The Lassa NPn along with Lassa NPc and several unrelated viral proteins were immobilized to unique sets of MagPlex microspheres using the standard immobilization protocol provided by the manufacturer.

Using a 10-fold excess of EZ-Link NHS-LC-LC-Biotin (Thermo Fisher Scientific), each sdAb was biotinylated for 30 min; excess biotin was removed using Zeba spin columns (Thermo Fisher Scientific). The absorbance at 280 nm was used to calculate the concentration of biotinylated sdAb (Bt-sdAb). Dilutions of each Bt-sdAb in PBSTB (phosphate-buffered saline [PBS] + 0.05% Tween + 0.1% bovine serum albumin) were prepared in round-bottom polypropylene microtiter plates (VWR). The mixture of antigen-coated MagPlex microspheres was added to the wells. The plate was washed and then incubated with 5 µg/mL streptavidin-conjugated phycoerythrin (SA-PE) for 30 min, washed, and binding evaluated on the MAGPIX instrument (Luminex).

Homogeneous sandwich format MagPlex bead assays were performed in order to demonstrate the ability of the sdAb to act as both the capture and recognition reagent for the detection of Lassa VLPs. For this assay, each sdAb was immobilized to a set of MagPlex microspheres as described above and each was then tested for their ability to function as a capture antibody. Initial tests evaluated all the clones as the biotinylated recognition molecule in the assay, however, as clone H6 provided the best response, the assay was repeated using only Bt-H6.

## 3. Results

After a llama was immunized with the Lassa NPn, a phage display library was generated to capture the VHH immune repertoire of the animal. Following three rounds of panning on LASV NPn recombinant protein, the evaluation of monoclonal phage from rounds two and three by MagPlex and ELISA generated 22 putative binding clones. Ten positive clones with signals at least twice the background were identified from the second round out of 96 screened, while 12 out of 96 third-round clones had signal at least four times the background and were chosen for further evaluation. Sequencing the selected prospective Lassa NPn binding sdAb revealed that the isolated clones sorted into four distinct sequence families based on homology of their CDR 3 sequence. One family was found to dominate with 16 members while the other families were represented by only one, two, or three clones. The large sequence family possessed an extra pair of cysteine residues that are presumed to form a noncanonical disulfide bond between Framework 2 and CDR 3 [21,22]; we term this the two disulfide bond (2-DSB) family. The 2-DSB family also had a relatively long CDR 3. The eight clones shown in Figure 1 represent clones from the four families that were prepared and evaluated. The G11 family had one other member, the E10 family had two other members, and E1 was the only member of its family.

```
          1       10        20        30        40        50        60        70        80        90        100       110       120       130  135
F3        DVQLQASGGGLVQPGGSLRLSCAASGFTLVYYGIGWFRQAPGKEREGVSCISSYDGTTYYADSVKGRFTISRDNAKNTVYLQMNSLKPEDTAVYYCATELGNTHRGSYFPKQRVERCHTGDFASWGQGTQVTVSS
E11       EVQLQASGGGLAQPGGSLRLSCAASGFTLVYYGIGWFRQAPGKEREGVSCISSNDGTTYYADSVKGRFTISRDNAKNTVYLQMNSLKPEDTAVYYCATELGNTHRGSYFPKQRVERCHTGDFASWGQGTQVTVSS
C7        EVQLQASGGGLVQPGGSLRLSCVASGFTLVYYGIGWFRQAPGKEREGVSCISSNDGSTYYADSVKGRFTISRDAAKNTMYLQMNSLKPEDTAVYYCATELGNTHRGSFFPNRRVERCRTDDFASWGQGTQVTVSS
B5        EVQLQASGGGLAQPGGSLRLSCAASGFTLVYYGIGWFRQAPGKEREGVSCISSNDGSTYYADSVKGRFTISRDAAKNTVYLQMNSLKPEDTAVYYCATELGNTHRGSFFPNRRVERCRTDDFASWGQGTQVTVSS
H6        DVQLQASGGGVVQPGGSLRLSCVASGFTLVYYGIGWFRQAPGKEREGVSCISSNDGSTYYADSVKGRFTISRDAAKNTVYLQMNSLKPEDTAVYYCATELGNTHRGSFFPNRRVKRCRTDDFASWGQGTQVTVSS
G11       DVQLQASGGGLVQAGDSLTLSCAASGRTVSPYGVGWFRQAPGREREFVAALSRSGTFTAYADSVKGRFTISRDNAKNTVYLQMNSLKPDDTAVYHCAARSAS----SYYSTYRKE----DDYHYWGQGTQVTVSS
E10       DVQLQASGGGLVQAGASLRLSCAASGFTYT---MGWFRQAPGQEREFVAAIARSGGTTYYTDSVKGRFTISRDNAKKTVFLQLNSLKPEDTAVYYCAADA---------LPYSAQSMSTRNPGYWGQGTQVTVSS
E1        DVQLQASGGGLVQTGGSLRLSCENSGRALRI--MGWFRQAPGKEREFVATINRNGESTYYSDSVKGRFTISRDNNKNTVYLQMNNLKPEDTGVYNCAADR---------PGPTQIKTRADDYDYWGEGTQVTVSS
Consensus #VQLQASGGGLvQ.GgSLrLSCaaSGftl..yg.GWFRQAPGkEREfVa.isr.gg.TyYaDSVKGRFTISRDnaKnTV%LQ$NsLKP#DTaVYyCAa........s..p..r.....td#..yWG#GTQVTVSS
```

**Figure 1.** Protein sequences of the eight representative sdAb that were produced and evaluated. Sequences are presented using one-letter amino acid abbreviations. Multalin was used to generate the alignment; red denotes high homology positions while lower homology is in blue [23]. Numbering is sequential based on the F3 sequence. Using this numbering, CDR 1 is from 26 to 33, CDR 2 is from 51 to 58, and CDR3 is from 97 to 124.

Five representative sequences from the 2-DSB family, and one each from the other families, were cloned into the pET22b expression vector and produced via *E. coli* in half-liter-scale shake flasks and purified. The protein produced was virtually all monomeric with very little aggregated material observed. Table 1 shows the protein yields. For these sdAb, the presence of the additional disulfide bond appeared to impair production, as the three clones having only the single consensus disulfide bond all produced at levels of 5 mg/L or better, while the 2-DSB family ranged from a low of 0.8 to a high of 3.5 mg/L. The addition of the pTUM4 helper plasmid [24], which we have shown to aid in the production of sdAb with multiple disulfide bonds [25], provided less than two-fold improvement in protein production, suggesting that it is possible another feature of the 2-DSB family negatively impacts production.

The melting temperature (Tm) and refolding ability of each clone were measured using CD (Table 1). The 2-DSB family had Tms that ranged from 63 to 66 °C, suggesting the point mutations within the family members had little impact on Tm. The refolding ability of the 2-DSB family members, however, varied from a low of 30% to a high of 93%.

**Table 1.** Production, CD melting point and refolding, and biophysical properties.

| sdAb | Yield (mg/L) | Tm (°C) | % Refold | pI | #SS Bonds |
|---|---|---|---|---|---|
| E1 | 8.4 | 59 | 56 | 7.18 | 1 |
| G11 | 5.2 | 60 | 30 | 8.01 | 1 |
| E10 | 16.4 | 48 | 15 | 8.64 | 1 |
| E11 | 0.8 | 64 | 93 | 7.95 | 2 |
| B5 | 1.4 | 63 | 52 | 7.96 | 2 |
| H6 | 3.5 | 64 | 30 | 8.74 | 2 |
| F3 | 2.7 | 65 | 59 | 7.89 | 2 |
| C7 | 1.9 | 66 | 83 | 7.96 | 2 |

Refolding ability showed no proportionality with production potential, suggesting no linkage between refolding and production. None of the other three families that had only the single conserved disulfide bond had a Tm over 60 °C or refolded remarkably well. Thus, exploring some possible sequence modifications that could lead to enhanced Tm and refolding ability for these binders, and perhaps codon optimization of the 2-DSB family to improve production, would be useful pursuits [26].

The binding kinetics of the sdAb on a surface coated with Lassa NPn recombinant protein were evaluated by SPR; results are presented in Table 2 (color represents different sequence families). Characteristic SPR plots are shown in Appendix A, Figure A1. The three families with only the canonical disulfide bond showed relatively poor affinities in comparison to the 2-DSB family, ranging from a KD of 69 nM to having no binding observed at all. The 2-DSB family all have superior off rates which ranged from 7.6 to 3.0 × $10^{-3}$ and the KD for the family was approximately 10 nM. None of the sdAb showed any binding to NPc (not shown).

**Table 2.** Kinetic parameters of Lassa NPn binding sdAb.

| Clone | ka (1/Ms) | kd (1/s) | KD (nM) |
|---|---|---|---|
| E1 | - | - | NBO * |
| G11 | $3.8 \times 10^5$ | $7.0 \times 10^{-2}$ | 184 |
| E10 | $1.9 \times 10^6$ | $1.3 \times 10^{-1}$ | 69 |
| E11 | $1.5 \times 10^5$ | $7.6 \times 10^{-3}$ | 57 |
| B5 | $3.2 \times 10^5$ | $6.1 \times 10^{-3}$ | 19 |
| H6 | $6.1 \times 10^5$ | $5.7 \times 10^{-3}$ | 9.3 |
| F3 | $4.0 \times 10^5$ | $3.0 \times 10^{-3}$ | 7.5 |
| C7 | $5.1 \times 10^5$ | $3.8 \times 10^{-3}$ | 7.5 |

* NBO: no binding observed.

The MagPlex direct binding assay of the Bt-sdAb to immobilized Lassa NPn confirmed the superiority of the 2-DSB family for their ability to bind relative to representatives of the three other families (Figure 2A). All sdAb pairs were examined for their ability to perform as both capture and recognition elements in a sandwich format. After some optimization and selection of the Bt-H6 sdAb as the best performing recognition antibody, we performed a homogeneous sandwich fluoroimmunoassay for the detection of Lassa VLPs (Figure 2B).

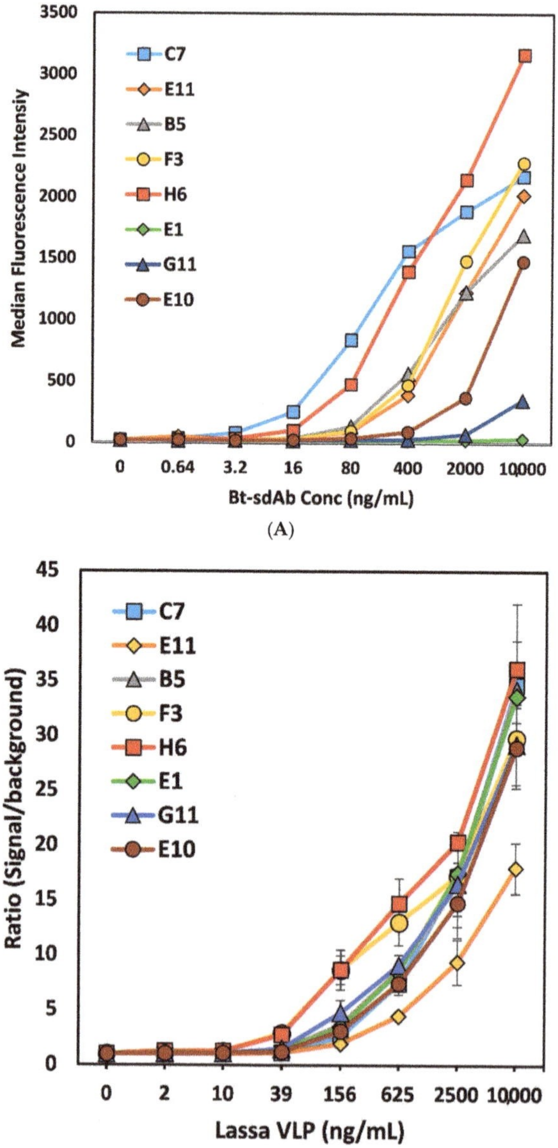


(**A**)

(**B**)

**Figure 2.** (**A**) MagPlex direct binding assay, where Lassa NPn was immobilized on MagPlex magnetic microspheres and biotinylated sdAb allowed to bind at different concentrations, followed by addition of SA-PE. (**B**) Homogeneous MagPlex sandwich fluoroimmunoassay; each sdAb was immobilized on a set of MagPlex magnetic microspheres and incubated in the presence of Lassa VLPs and biotinylated sdAb H6 and SA-PE. The amount of SA-PE bound to each microsphere set was evaluated following a single wash step. This experiment was repeated on three separate days in triplicate. Shown is the average of the three replicates from one experiment; the error bars represent the standard deviation of the mean.

## 4. Discussion

This work describes the initial selection and characterization of sdAb that are specific for the Lassa NPn recombinant protein. Most of the selected binders fell into one large family where all the members possessed a second disulfide bond between framework 2 and CDR 3, typical of subfamily 3 of llama VHH [22]. As with other members of subfamily 3, the isolated 2-DSB family members possess a longer-than-average CDR3 with a length of 27 versus conventional llama sdAb whose average CDR 3 length is 14 [22]. The sdAb's conserved disulfide bond primarily provides enhanced stability to the structure of the sdAb, while the disulfide bond between framework 2 and CDR 3 can stabilize the structure; it typically plays a more critical role in binding affinity by maintaining the secondary structure of the binding loops [27,28]. Three other sequence families were also identified which had between one and three representatives. These three other clones had only the typical disulfide bond found in variable heavy domains, however, they still had long CDR 3 regions, with lengths of 18 and 19.

The melting point and ability to refold were evaluated for each of the clones. None of them were found to be remarkably stable although clone E11 did refold over 90%. It may be of interest to better understand the impact that the minor sequence differences between members of the 2-DSB family have on their ability to refold as it varied from a low of 30% for H6 to a high of 93% for E11, while their Tm was virtually the same. If these binders are developed further, one could make additional sequence changes to increase both their Tm and ability to refold [26], thus making them more attractive for applications in austere locales where refrigeration is a challenge.

The affinity of each of these binders was also evaluated. The clone E1 did not show any binding by SPR or give a significant signal in the MagPlex direct binding assay, thus it is likely that isolation of this sequence was spurious. Notably, E1 was the only member of its sequence family, increasing the likelihood that it is not a true Lassa NPn binder. Of the other clones, it quickly became clear that the 2-DSB family of binders were superior to the other two. Their affinities were in the range of 10 nM and all had off rates that were at least fivefold lower than any of the other binders examined. Nonetheless, these affinities are not as good as those frequently obtained from llama immune phage-display libraries, where frequently sub-nM KD binders are isolated. This limited affinity is borne out by the need to perform a homogenous sandwich assay, whereby limiting the dissociation time of the immune-complex, we were able to demonstrate that these binders show potential for the detection of LASV via their ability to detect Lassa VLPs that included NP in addition to glycoprotein and matrix protein.

After testing a range of conditions, we were able to demonstrate detection of Lassa VLPs down to as little as 39 ng/mL, with both the H6 and F3 capture sets having a ratio of signal/background over 3, where the limit of detection cut-off ratio is 2, a ratio we have shown previously provides a statistically significant increase [16]. While this demonstrates their potential, it is highly likely that binders with a higher affinity will be required to obtain adequate sensitivities to reliably detect for LASV infection. To achieve this goal, the two best options are to immunize another animal in an effort to obtain an immune repertoire that possesses higher affinity binders or to take advantage of avidity by preparing sdAb multimers of our current binders. In other work we, as well as others, have made sdAb multimers either via direct genetic linkage [29–31] or via a fusion protein that forms a homodimer such as alkaline phosphatase or rhizavidin [32,33] or fusion with a peptide that promotes multimerization [34]. These dimers and multimers have been shown to dramatically enhance both affinity and utility in immunoassays [35,36], however, it remains to be determined if such an approach would prove successful here.

The sdAb were selected on a recombinantly expressed n-terminal fragment of NP containing amino acids 1-304 of the protein from the lineage IV strain Josiah. A comparison of NP sequences from different LASV strains showed that there can be up to 12% variation in the amino acid sequence [4]. Future work should include determining if the selected sdAb are able to recognize NP from other LASV strains. Alternatively, one could do additional rounds of selection with a different LASV strain to select binders capable of binding to a common epitope. Nonetheless, the sdAb we isolated could be used in conjunction with sdAb reagents selected on NP from a variety of strains to minimize false negatives.

Future studies involve evaluating the ability of the sdAb to recognize VLPs that include NP derived from other LASV lineages, and eventually their ability to detect live virus. Ultimately, one will want to compare sdAb reagents head-to-head with conventional monoclonal antibodies to directly evaluate any perceived advantage obtained and perhaps whether an assay format that includes both reagents is warranted. However, from our experience, until the sdAb have been further optimized to enhance their functionality, such a comparison would be premature. In conclusion, we have described and evaluated four families of Lassa NPn binding sdAb and hypothesize that with continued development, they may become useful reagents in the fight against Lassa fever.

**Author Contributions:** Investigation: J.L.L., E.R.G., G.P.A. and L.C.S.-L.; Project administration: G.P.A. and E.R.G.; Resources: L.C.S.-L.; Supervision: G.P.A. and E.R.G.; Visualization: G.P.A., L.C.S.-L. and E.R.G.; Writing—original draft: G.P.A. and E.R.G.; Writing—review & editing: G.P.A., L.C.S.-L., J.L.L. and E.R.G. All authors have read and agreed to the published version of the manuscript.

**Funding:** This work was supported by Naval Research Laboratory base funds.

**Acknowledgments:** The authors would like to acknowledge the support of Dan Zabetakis for his many suggestions and insightful discussions. They would also like to thank Luis Branco and Zalgene for helpful discussion and reagents.

**Conflicts of Interest:** The authors declare no conflict of interest. The funding sponsors had no role in the design of the study; in the collection, analyses, or interpretation of data; in the writing of the manuscript, and in the decision to publish the results.

## Appendix A

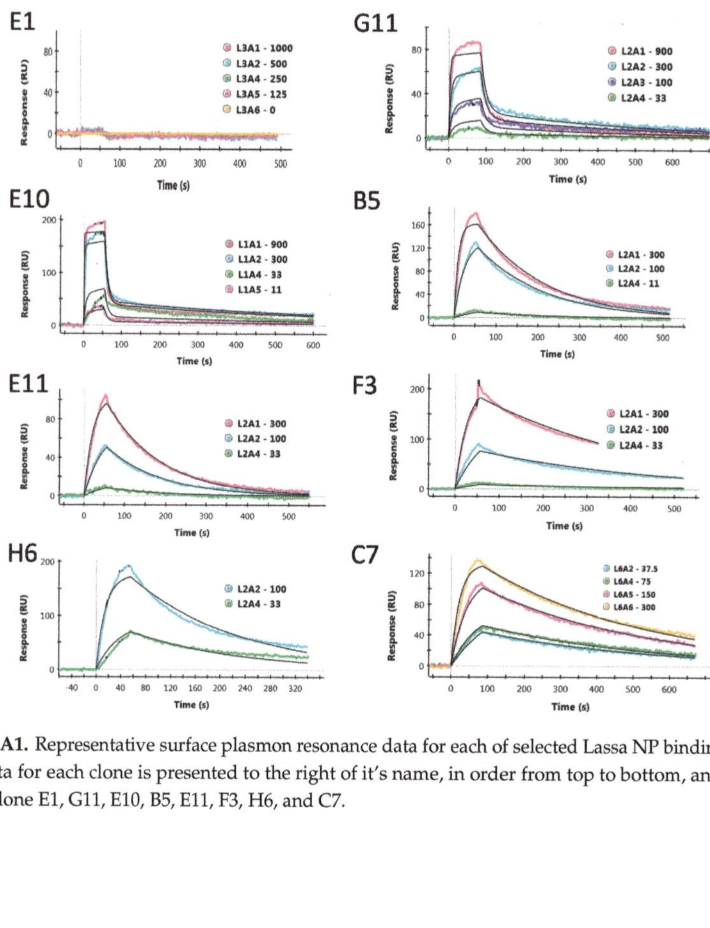


**Figure A1.** Representative surface plasmon resonance data for each of selected Lassa NP binding sdAb. The data for each clone is presented to the right of it's name, in order from top to bottom, and left to right, clone E1, G11, E10, B5, E11, F3, H6, and C7.

## References

1. Yun, N.E.; Walker, D.H. Pathogenesis of Lassa fever. *Viruses* **2012**, *4*, 2031–2048. [CrossRef] [PubMed]
2. Branco, L.M.; Boisen, M.L.; Andersen, K.G.; Grove, J.N.; Moses, L.M.; Muncy, I.J.; Henderson, L.A.; Schieffellin, J.S.; Robinson, J.E.; Bangura, J.J.; et al. Lassa hemorrhagic fever in a late term pregnancy from northern sierra leone with a positive maternal outcome: Case report. *Virol. J.* **2011**, *8*, 404. [CrossRef] [PubMed]
3. Price, M.E.; Fisher-Hoch, S.P.; Craven, R.B.; McCormick, J.B. A prospective study of maternal and fetal outcome in acute Lassa fever infection during pregnancy. *BMJ* **1988**, *297*, 584–587. [CrossRef] [PubMed]
4. Bowen, M.D.; Rollin, P.E.; Ksiazek, T.G.; Hustad, H.L.; Bausch, D.G.; Demby, A.H.; Bajani, M.D.; Peters, C.J.; Nichol, S.T. Genetic Diversity among Lassa Virus Strains. *J. Virol.* **2000**, *74*, 6992–7004. [CrossRef] [PubMed]
5. Raabe, V.; Koehler, J. Laboratory Diagnosis of Lassa Fever. *J. Clin. Microbiol.* **2017**, *55*, 1629–1637. [CrossRef] [PubMed]
6. Boisen, M.L.; Hartnett, J.N.; Shaffer, J.G.; Goba, A.; Momoh, M.; Sandi, J.D.; Fullah, M.; Nelson, D.K.S.; Bush, D.J.; Rowland, M.M.; et al. Field validation of recombinant antigen immunoassays for diagnosis of Lassa fever. *Sci. Rep.* **2018**, *8*, 5939. [CrossRef]
7. Boisen, M.L.; Uyigue, E.; Aiyepada, J.; Siddle, K.J.; Oestereich, L.; Nelson, D.K.S.; Bush, D.J.; Rowland, M.M.; Heinrich, M.L.; Eromon, P.; et al. Field evaluation of a Pan-Lassa rapid diagnostic test during the 2018 Nigerian Lassa fever outbreak. *Sci. Rep.* **2020**, *10*, 8724. [CrossRef]
8. Jahrling, P.; Niklasson, B.; McCormick, J. Early Diagnosis of Human Lassa Fever by ELISA Detection of Antigen and Antibody. *Lancet* **1985**, *325*, 250–252. [CrossRef]
9. Bausch, D.G.; Rollin, P.E.; Demby, A.H.; Coulibaly, M.; Kanu, J.; Conteh, A.S.; Wagoner, K.D.; McMullan, L.K.; Bowen, M.D.; Peters, C.J.; et al. Diagnosis and clinical virology of Lassa fever as evaluated by enzyme-linked immunosorbent assay, indirect fluorescent-antibody test, and virus isolation. *J. Clin. Microbiol.* **2000**, *38*, 2670–2677. [CrossRef]
10. Branco, L.M.; Grove, J.N.; Boisen, M.L.; Shaffer, J.G.; Goba, A.; Fullah, M.; Momoh, M.; Grant, D.S.; Garry, R.F. Emerging trends in Lassa fever: Redefining the role of immunoglobulin M and inflammation in diagnosing acute infection. *Virol. J.* **2011**, *8*, 478. [CrossRef]
11. Grove, J.N.; Branco, L.M.; Boisen, M.L.; Muncy, I.J.; Henderson, L.A.; Schieffellin, J.S.; Robinson, J.E.; Bangura, J.J.; Fonnie, M.; Schoepp, R.J.; et al. Capacity building permitting comprehensive monitoring of a severe case of Lassa hemorrhagic fever in Sierra Leone with a positive outcome: Case report. *Virol. J.* **2011**, *8*, 314. [CrossRef] [PubMed]
12. Hamers-Casterman, C.; Atarhouch, T.; Muyldermans, S.; Robinson, G.; Hamers, C.; Songa, E.B.; Bendahman, N.; Hamers, R. Naturally occurring antibodies devoid of light chains. *Nature* **1993**, *363*, 446–448. [CrossRef] [PubMed]
13. Muyldermans, S. Nanobodies: Natural Single-Domain Antibodies. *Annu. Rev. Biochem.* **2013**, *82*, 775–797. [CrossRef] [PubMed]
14. de Marco, A. Recombinant expression of nanobodies and nanobody-derived immunoreagents. *Protein Expr. Purif.* **2020**, *172*, 105645. [CrossRef]
15. Liu, J.L.; Shriver-Lake, L.C.; Anderson, G.P.; Zabetakis, D.; Goldman, E.R. Selection, characterization, and thermal stabilization of llama single domain antibodies towards Ebola virus glycoprotein. *Microb. Cell Fact.* **2017**, *16*, 223. [CrossRef]
16. Shriver-Lake, L.C.; Liu, J.L.; Zabetakis, D.; Sugiharto, V.A.; Lee, C.-R.; Defang, G.N.; Wu, S.-J.L.; Anderson, G.P.; Goldman, E.R. Selection and Characterization of Anti-Dengue NS1 Single Domain Antibodies. *Sci. Rep.* **2018**, *8*, 1–11. [CrossRef]
17. Goldman, E.R.; Anderson, G.P.; Liu, J.L.; Delehanty, J.B.; Sherwood, L.J.; Osborn, L.E.; Cummins, L.B.; Hayhurst, A. Facile generation of heat-stable antiviral and antitoxin single domain antibodies from a semisynthetic llama library. *Anal. Chem.* **2006**, *78*, 8245–8255. [CrossRef]
18. Walper, S.A.; Liu, J.L.; Zabetakis, D.; Anderson, G.P.; Goldman, E.R. Development and Evaluation of Single Domain Antibodies for Vaccinia and the L1 Antigen. *PLoS ONE* **2014**, *9*, e106263. [CrossRef]
19. Shriver-Lake, L.C.; Zabetakis, D.; Goldman, E.R.; Anderson, G.P. Evaluation of anti-botulinum neurotoxin single domain antibodies with additional optimization for improved production and stability. *Toxicon* **2017**, *135*, 51–58. [CrossRef]

20. Walper, S.A.; Lee, P.A.B.; Goldman, E.R.; Anderson, G.P. Comparison of single domain antibody immobilization strategies evaluated by surface plasmon resonance. *J. Immunol. Methods* **2013**, *388*, 68–77. [CrossRef]
21. Muyldermans, S.; Atarhouch, T.; Saldanha, J.; Barbosa, J.A.; Hamers, R. Sequence and structure of VH domain from naturally occurring camel heavy chain immunoglobulins lacking light chains. *Protein Eng.* **1994**, *7*, 1129–1135. [CrossRef] [PubMed]
22. Harmsen, M.M.; Ruuls, R.C.; Nijman, I.J.; Niewold, T.A.; Frenken, L.G.J.; de Geus, B. Llama heavy-chain V regions consist of at least four distinct subfamilies revealing novel sequence features. *Mol. Immunol.* **2000**, *37*, 579–590. [CrossRef]
23. Corpet, F. Multiple sequence alignment with hierarchical-clustering. *Nucleic Acids Res.* **1988**, *16*, 10881–10890. [CrossRef] [PubMed]
24. Schlapschy, M.; Grimm, S.; Skerra, A. A system for concomitant overexpression of four periplasmic folding catalysts to improve secretory protein production in Escherichia coli. *Protein Eng. Des. Sel.* **2006**, *19*, 385–390. [CrossRef]
25. Shriver-Lake, L.C.; Goldman, E.R.; Zabetakis, D.; Anderson, G.P. Improved production of single domain antibodies with two disulfide bonds by co-expression of chaperone proteins in the Escherichia coli periplasm. *J. Immunol. Methods* **2017**, *443*, 64–67. [CrossRef]
26. Goldman, E.R.; Liu, J.L.; Zabetakis, D.; Anderson, G.P. Enhancing Stability of Camelid and Shark Single Domain Antibodies: An Overview. *Front. Immunol.* **2017**, *8*, 865. [CrossRef]
27. Hagihara, Y.; Mine, S.; Uegaki, K. Stabilization of an immunoglobulin fold domain by an engineered disulfide bond at the buried hydrophobic region. *J. Biol. Chem.* **2007**, *282*, 36489–36495. [CrossRef]
28. Govaert, J.; Pellis, M.; Deschacht, N.; Vincke, C.; Conrath, K.; Muyldermans, S.; Saerens, D. Dual Beneficial Effect of Interloop Disulfide Bond for Single Domain Antibody Fragments. *J. Biol. Chem.* **2012**, *287*, 1970–1979. [CrossRef]
29. Goldman, E.R.; Broussard, A.; Anderson, G.P.; Liu, J.L. Bglbrick strategy for the construction of single domain antibody fusions. *Heliyon* **2017**, *3*, e00474. [CrossRef]
30. Krasniqi, A.; Bialkowska, M.; Xavier, C.; Van der Jeught, K.; Muyldermans, S.; Devoogdt, N.; D'Huyvetter, M. Pharmacokinetics of radiolabeled dimeric sdAbs constructs targeting human CD20. *New Biotechnol.* **2018**, *45*, 69–79. [CrossRef]
31. Conrath, K.E.; Lauwereys, M.; Wyns, L.; Muyldermans, S. Camel single-domain antibodies as modular building units in bispecific and bivalent antibody constructs. *J. Biol. Chem.* **2001**, *276*, 7346–7350. [CrossRef] [PubMed]
32. Liu, J.L.; Zabetakis, D.; Brozozog Lee, P.A.; Goldman, E.R.; Anderson, G.P. Single Domain Antibody Alkaline Phosphatase Fusion Proteins for Antigen Detection—Analysis of Affinity and Thermal Stability of Single Domain Antibody. *J. Immunol. Methods* **2013**, *393*, 1–7. [CrossRef] [PubMed]
33. Liu, J.L.; Zabetakis, D.; Walper, S.A.; Goldman, E.R.; Anderson, G.P. Bioconjugates of rhizavidin with single domain antibodies as bifunctional immunoreagents. *J. Immunol. Methods* **2014**, *411*, 37–42. [CrossRef]
34. Zhang, J.; Tanha, J.; Hirama, T.; Khieu, N.H.; To, R.; Tong-Sevinc, H.; Stone, E.; Brisson, J.-R.; MacKenzie, C.R. Pentamerization of Single-domain Antibodies from Phage Libraries: A Novel Strategy for the Rapid Generation of High-avidity Antibody Reagents. *J. Mol. Biol.* **2004**, *335*, 49–56. [CrossRef]
35. Swain, M.D.; Anderson, G.P.; Serrano-Gonzalez, J.; Liu, J.L.; Zabetakis, D.; Goldman, E.R. Immunodiagnostic reagents using llama single domain antibody-alkaline phosphatase fusion proteins. *Anal. Biochem.* **2011**, *417*, 188–194. [CrossRef]
36. Wang, S.; Zheng, C.; Liu, Y.; Zheng, H.; Wang, Z. Construction of multiform scFv antibodies using linker peptide. *J. Genet. Genom.* **2008**, *35*, 313–316. [CrossRef]

antibodies

MDPI

Article

# Crystal Structure and Characterization of Human Heavy-Chain Only Antibodies Reveals a Novel, Stable Dimeric Structure Similar to Monoclonal Antibodies

**Carl Mieczkowski [1,*,†], Soheila Bahmanjah [2,†], Yao Yu [1], Jeanne Baker [1], Gopalan Raghunathan [1], Daniela Tomazela [1], Mark Hsieh [1], Mark McCoy [3], Corey Strickland [2] and Laurence Fayadat-Dilman [1]**

1 Discovery Biologics, Protein Sciences, Merck & Co., Inc., South San Francisco, CA 94080, USA; yao.yu@merck.com (Y.Y.); jeanne.baker@merck.com (J.B.); gopalan.raghunathan@merck.com (G.R.); daniela.tomazela@merck.com (D.T.); mark_hsieh@merck.com (M.H.); laurence.fayadat-dilman@merck.com (L.F.-D.)

2 Department of Chemistry, Modeling and Informatics, Merck & Co., Inc., Kenilworth, NJ 07033, USA; soheila.bahmanjah@merck.com (S.B.); corey.strickland@merck.com (C.S.)

3 Department of Pharmacology, Mass Spectrometry & Biophysics, Merck & Co., Inc., Kenilworth, NJ 07033, USA; mark.mccoy@merck.com

* Correspondence: carl.mieczkowski@merck.com; Tel.: +1-650-496-6501

† Both authors contributed equally to this work.

Received: 12 September 2020; Accepted: 9 November 2020; Published: 22 November 2020

**Abstract:** We report the novel crystal structure and characterization of symmetrical, homodimeric humanized heavy-chain-only antibodies or dimers (HC2s). HC2s were found to be significantly coexpressed and secreted along with mAbs from transient CHO HC/LC cotransfection, resulting in an unacceptable mAb developability attribute. Expression of full-length HC2s in the absence of LC followed by purification resulted in HC2s with high purity and thermal stability similar to conventional mAbs. The $V_H$ and $C_H1$ portion of the heavy chain (or Fd) was also efficiently expressed and yielded a stable, covalent, and reducible dimer (Fd2). Mutagenesis of all heavy chain cysteines involved in disulfide bond formation revealed that Fd2 intermolecular disulfide formation was similar to Fabs and elucidated requirements for Fd2 folding and expression. For one HC2, we solved the crystal structure of the Fd2 domain to 2.9 Å, revealing a highly symmetrical homodimer that is structurally similar to Fabs and is mediated by conserved ($C_H1$) and variable ($V_H$) contacts with all CDRs positioned outward for target binding. Interfacial dimer contacts revealed by the crystal structure were mutated for two HC2s and were found to dramatically affect HC2 formation while maintaining mAb bioactivity, offering a potential means to modulate novel HC2 formation through engineering. These findings indicate that human heavy-chain dimers can be secreted efficiently in the absence of light chains, may show good physicochemical properties and stability, are structurally similar to Fabs, offer insights into their mechanism of formation, and may be amenable as a novel therapeutic modality.

**Keywords:** antibody; heavy-chain dimer; heavy-chain antibody; crystal structure

## 1. Introduction

Typical monoclonal antibodies (mAbs) with specificity towards a target antigen are composed of heavy (HC) and light (LC) chains containing conserved and variable regions. Previously, heavy-chain only antibody (HCAb) formation was reported to occur in various species with significant human therapeutic potential [1]. Camelids are long known to express functional HC-only antibodies that are composed of a homodimeric $V_{HH}$ domain [2,3]. Further, sharks produce functional heavy-chain only antibodies, that like camelid antibodies, are smaller in nature, and formed the basis of nanobody technology [4,5]. Like camelid $V_{HH}$ domains and shark nanobodies, both lacking $C_H1$ and LC domains, HCAbs have been reported to be secreted in LC-deficient mice lacking the $C_H1$ domain [6]. Separately, hybrid llama/human antibody HCAbs, lacking the $C_H1$ domain and having swapped the llama $V_{HH}$ regions with human $V_H$, have been reported [7]. In addition, HC-only transcripts, lacking the $C_H1$ domain and in the absence of LC, can be expressed on the cell surface of mammalian pro-B cells [8]. What is noteworthy here with these examples of HC-only antibodies found in camelids, sharks, LC-deficient mice, and mammalian pro-B cells is that the presence of these molecules does not contradict the longstanding views on antibody mAb or Fab assembly, where LC assembly to the HC, or in particular to the $C_H1$ domain, is required for $C_H1$ domain folding and dissociation from the molecular chaperone BiP [9–11]. Interestingly, it has been reported that full- length HC-only antibody dimers are formed from a stable Drosophila cell line via a BiP mediated pathway [12]. This observation challenges the long-held hypothesis that the unfolded $C_H1$ domain in complex to the molecular chaperone BiP requires association with LC to fold and release BiP chaperone, enabling export and secretion. Nonetheless, the formation of full-length HC-only antibodies is uncommon, and aside from the normal requirement of the LC to bind chaperoned $C_H1$ and release BiP, additional mechanisms may be required to neutralize their potential toxicity in the absence of LC as previously reported in plasma cells [13].

Human HCAbs have only recently been reported by Stoyle and coworkers to occur from transient Chinese Hamster Ovary (CHO) expression [14]. Like antibody producing B cells, CHO cells have a similar quality control system and mechanism of antibody assembly, utilizing BiP, prolyl isomerases, and disulfide reductases [15]. Therein, HCAbs containing the constant $C_H1$ and $V_H$ regions humanized from rodent sources were found to form homodimers and be secreted even in the absence of light chain. These HC dimers were found to form from both HC/LC cotransfected cells and HC-only transfection, and both full-length HC dimers and HC dimers lacking the Fc domain ($V_H$ + $C_H1$ only) were able to form. The LC-independent secretion of HC dimers was inferred to be variable region dependent since only certain HCs were able to form and be secreted as folded molecules. One characteristic noted for some of the molecules being able to form HC2s was the increased number of positively charged amino acids in the HC-CDR3 [14].

Although Stoyle and coworkers clearly showed that both full-length heavy-chain dimers (HC2s) and HCs lacking the Fc domain (analogous to the Fab domain, herein described as "Fd2") are able to be formed and secreted in the absence of LC, it is not clear what their mechanism of assembly is, nor is it clear what are the sequence and structural determinants that drive their formation. Also lacking are biophysical and structural characterizations of these novel molecules. The formation of human HC2s is unique and offers extraordinary insights into antibody assembly, as well as potentially enabling different, novel antibody formats with unique advantages and properties. One potential advantage is that common LC mispairing in multi-specific formats can be avoided. However, without further understanding their biophysical properties and how structural and sequence elements impact their formation, engineering and modulating HC dimers into a novel potential modality cannot be achieved. Aside from assumed differences in the HC variable sequence that impact HC2 formation, the role of the LC is also not well understood and how this competing pathway to normal HC-LC assembly could impact HC-HC (HC2) assembly.

Herein, we report robust expression and formation of human HC2s for two different humanized antibodies with unrelated variable sequences against different biological targets. Robust formation of HC2 impurities originally constituted a severe, negative mAb developability attribute. In HC/LC cotransfected transient CHO cells, HC2 formation was robust and prevented further developability of these clones as acceptable purity/heterogeneity of the desired mAb species was unattainable under representative process conditions. To characterize further these novel and unique "impurities", we performed a series of HC-only transfections and characterized the secreted and purified HC dimers. For both full-length and Fd2 versions, we found these molecules can achieve high purity, and to have similar or better expression titer, thermal stability, and accelerated storage stability than their mAb and Fab counterparts, respectively. Additionally, we solved the first ever crystal structure at 2.9 Angstroms (Å) of a human HC homodimer (herein named Fd2-A) that reveals exposed CDR regions and a symmetrical dimerization complex analogous to HC-LC association in Fabs, where one opposing HC ($C_H1$ + $V_H$ or Fd) replaces the LC. Disulfide formation is overall conserved between this Fd2 complex in comparison to the Fd domain in Fabs. Mutagenesis of key, variable CDR residues in the dimerization interface indicated by the novel Fd2-A crystal structure reduced full-length HC dimer formation. Similar mutations were made for a second recombinant antibody based on a sequence alignment to Fd2-A and similar HC2 dimer reduction was achieved, suggesting that this heavy-chain only dimer structure is conserved for two antibody sequences with different germlines against two different biological targets. These results indicate that we have solved a novel crystal structure representing the human HC dimer (Fd2) structure, provided new insights into their formation, structural and sequence determinants, and demonstrated that their application as a novel and functional antibody format is realistic.

## 2. Materials and Methods

### 2.1. Transient Protein Expression

For the small-scale protein production, transient transfections were done in TubeSpin® bioreactors (TPP Techno Plastic Products AG, Trasadingen, Switzerland) using the ExpiCHO Expression System (Thermo Fisher Scientific, Waltham, MA, USA) according to the manufacturer's protocol. Briefly, the cells were grown and maintained in ExpiCHO Expression Medium (Thermo Fisher Scientific, Waltham, MA, USA) and seeded in 10 mL of media at $6 \times 10^6$ cells/mL on the day of transfection. Complexes were formed with 8 ug of DNA and 32 μL of Expifectamine in OptiPRO™ SFM and incubated for 1 min followed by addition to the cells. The transfected cultures were grown at 37 °C, 5% $CO_2$, 80% humidity, and 300 rpm rotation in a Multitron incubator (Infors HT, Basel, Switzerland) and then shifted to 32 °C 24 h post-transfection and were fed with feed and enhancer on days 1 and 5. Expression variables include no enhancer treatment (Thermo), pulling culture on Day 4, using 20% of coding DNA by weight, transfecting with 5X LC by molar mass, and using a 10 mL shake flask instead of tubespin cultures. The cultures were harvested on day 7, the cells were pelleted by centrifugation, and the supernatant was passed through a 0.2 micron filter. The protein titers were determined using a ForteBio Octet (Molecular Devices, LLC. San Jose, CA, USA) with Protein A sensors and a purified mAb to generate the standard curve.

For large-scale protein production, transient transfections were performed in 1 L shake flasks using the ExpiCHO Expression System (Thermo Fisher Scientific, Waltham, MA, USA) according to the manufacturer's protocol. Transfections and expression conditions were the same as in small-scale format. The cultures were harvested on day 7, the cells were pelleted by centrifugation, and the supernatant was passed through a 0.2 micron filter. The protein titers were determined using a ForteBio Octet (Molecular Devices, LLC. San Jose, CA, USA) with Protein A sensors and a purified mAb to generate the standard curve.

### 2.2. Small-Scale Protein Purification

The clarified cell culture supernatants were loaded onto a Tecan Freedom EVO 200 (Tecan Life Sciences, Männedorf, Switzerland) for antibody purification utilizing miniature columns manufactured by Repligen (Waltham, MA, USA) and packed with MabSelect™ SuRe™ LX (GE Healthcare Life Sciences, Pittsburgh, PA, USA). The antibodies were eluted with 20 mM sodium acetate at pH 3.5 and immediately neutralized with 0.33 M Tris, 1 M sodium acetate pH 8.0 and buffer exchanged into 20 mM sodium acetate pH 5.5 using 10K MWCO Slide-A-Lyzer dialysis cassettes (Thermo Fisher Scientific, Waltham, MA, USA).

### 2.3. *Large-Scale mAb and Full-Length HC Dimer Purification*

Cell culture supernatant was incubated with Protein A (ProA) affinity MabSelect SuRe LX resin (GE Healthcare, Pittsburgh, PA, USA) overnight (1 mL resin for 40 mg target antibody or HC dimer estimated by ForteBio Octet) for batch binding. Affinity resin with bound protein was collected by filtration and transferred to a disposable column. Bound resin was washed with 20X column volumes of 1X Gibco Phosphate buffered saline (PBS) (Thermo Fisher Scientific, Waltham, MA, USA) in batch mode. Desired protein was eluted by approximately 5 column volumes of 20 mM sodium acetate pH 3.5 buffer. Eluate was immediately buffer exchanged into 20 mM sodium acetate pH 5.5 buffer using 10K MWCO Slide-A-Lyzer dialysis cassettes. To further polish material for analytical and biophysical characterization efforts (DSC, SEC-MALS, LC-MS), desired product was purified to >98% purity (by SE-UPLC and cSDS) on a Superdex 200 Increase 10/300GL column (GE Healthcare Bio-Science AB, Uppsala, Sweden). Mobile phase was 20 mM sodium acetate, 200 mM sodium chloride pH 5.5 buffer. All purified protein was buffer exchanged overnight using 10K MWCO Slide-A-Lyzer dialysis cassettes into 1X Gibco PBS pH 7.4 and normalized to 1 mg/mL for characterization. Concentration was determined by UV absorbance at 280 nm on a Nanodrop 2000 1- position Spectrophotometer (Thermo Scientific).

### 2.4. *Fab and Fd Dimer Purification*

Cell culture supernatant was incubated with CaptureSelect IgG-$C_H1$ affinity matrix (Thermo Scientific) overnight for batch binding (1 mL resin for 10 mg protein estimated by ForteBio Octet). Resin was collected by filtration, transferred to a disposable column, and then washed with 20 column volumes of 1X Gibco PBS in batch mode. Desired protein was eluted with 5 column volumes of 0.1 M glycine pH 3.0 buffer. Eluate was immediately buffer exchanged into 20 mM sodium acetate pH 5.5 buffer. To further polish material for biophysical characterization and crystallization efforts, desired product was purified to >98% purity (by SE-UPLC and cSDS) on a Superdex 200 increase 10/300GL (GE Healthcare Bio-Science AB, Uppsala, Sweden) column. Mobile phase was 20 mM sodium acetate, 200 mM sodium chloride pH 5.5 buffer. All purified protein was buffer exchanged overnight into 1X Gibco PBS pH 7.4 and normalized to 1 mg/mL for characterization. For crystallization efforts, Fd2 for Molecule A was reformulated into 20 mM sodium acetate pH 5.5 (low salt) and concentrated to 20 mg/mL using a 4 mL Vivaspin™ ultrafiltration spin column with 10K MWCO membrane. Purity was verified to be unchanged following concentration by SE-UPLC.

### 2.5. *Size-Exclusion Ultra-Pressure Liquid Chromatography (SE-UPLC)*

A Waters Acquity UPLC H-Class PLUS system (Waters, Milford, MA, USA) was used to separate molecules based on differences in their hydrodynamic size. Samples (10 μg) were injected into an Acquity BEH200 SEC column (Waters, Milford, MA, USA) and eluted at a 0.5 mL/min flow rate. Mobile phase contained 100 mM sodium phosphate and 200 mM sodium chloride pH 7.0. Waters BEH200 SEC protein standard mix (Waters, Milford, MA, USA) was used as molecular weight marker and for column quality control purposes. Samples were detected by UV absorbance at 280 nm. Chromatograms were integrated manually and reported as a % of integrated area for each species.

### 2.6. *Capillary Sodium Dodecyl-Sulfate Electrophoresis (cSDS)*

A LabChip GXII Clipper (Perkin Elmer, Waltham, MA, USA) was used to determine purity and approximate molecular weight under non-reduced and reduced conditions. All sample and chip preparation was done according to manufacturer's protocol using the Protein Express Assay reagent kit (Perkin Elmer). Five μL of sample was mixed with 35 μL of reduced Protein Express Sample Buffer (35 mM Dithiothreitol or DTT added to kit sample buffer) or 35 μL non-reduced sample buffer (35 mM iodoacetamide added to kit sample buffer). An HT Protein Express Assay LabChip (Perkin Elmer) was primed using the Protein Express Assay Reagent kit (Perkin Elmer). LabChip GX Reviewer software was used for data analysis. Chromatograms were integrated manually and reported as % for each species. This method produces system or method-related peaks that appear at before 10 KDa in size and include reagents and the internal 10 KDa molecular weight standard.

### 2.7. *Size Exclusion Chromatography Coupled to Multi-Angle Light Scattering (SEC-MALS)*

A µDAWN (Wyatt Technology, Santa Barbara, CA, USA) was coupled online to a Waters UPLC H-Class system (Waters, Milford, MA, USA) to measure molecular weight (Mw) using static light scattering. Light scattering wavelength used was 650 nm. The concentration detector was an online uTrex equipped with a refractive index detector. Dn/Dc value used in Mw calculations was 0.185 mL/g. PBS was used as the mobile phase and 20 µg samples were injected. Other SE-UPLC running conditions used are described here in the SE-UPLC methods section. Bovine serum albumin (Thermo Scientific) was used as an isotropic standard for molecular weight normalization. Astra 7 software (Wyatt Technology, Santa Barbara, CA, USA) was used for data acquisition and analysis. Detector alignment, band broadening, and normalization coefficient parameters were set to achieve a BSA Mw within 5% of 66,500 Da.

### 2.8. *Differential Scanning Calorimetry (DSC)*

A MicroCal PEAQ-DSC (Malvern Panalytical, Malvern, UK) was used to measure protein melting temperatures. An amount of 500 µL of protein solution for each sample at 1 mg/mL in 20 mM sodium acetate pH 5.5 was added to a 96-well 500 µL volume plate (Wheaton). Temperature was ramped from 25 °C to 95 °C at 1 °C per min. Origin 7 software was used for data acquisition and analysis where melting transition temperatures ($T_m$) were fitted using a non-two state algorithm.

### 2.9. *Nanoscale-Differential Scanning Fluorimetry (Nano-DSF)*

All nano-DSF studies were performed using the Nanotemper Prometheus NT.48 instrument and data analysis software. Samples were evaluated at 1 mg/mL in 20 mM sodium acetate pH 5.5. Samples were introduced by capillary action into glass capillaries (Prometheus) prior to placing into the instrument capillary holder. Temperature was ramped from 20 °C to 95 °C at 1 °C/minute. Thermal melting transitions ($T_m$) were measured by identifying changes in inflection of intrinsic fluorescence intensity ratios (F350 nm/F330 nm) at specific temperatures.

### 2.10. *Surface Plasmon Resonance (SPR) Affinity Measurements by BIAcore*

Binding kinetics of the mAbs to the target was determined by SPR on a BIAcore T200 (GE Healthcare, Chicago, IL, USA). The running buffer, 10 mM HEPES, 150 mM NaCl, 0.05% (*v*/*v*) Surfactant P20, 3 mM EDTA, pH 7.4 (HBS-EP+, GE Healthcare), was used for immobilization and reagent dilutions. All binding kinetics were measured at 25 °C. For each injection cycle, mAbs were first captured in different flow cells with an anti-human Fc antibody (Human Antibody Capture Kit, GE Healthcare) immobilized to the sensor chip (Series S CM5, GE Healthcare). Reference flow cell with no captured mAb was also used. Serial dilutions (1:2) of the target protein, ranging in concentration from 1 µM to 32 µM, and buffer blanks were injected in multiple cycles over the captured mAbs and reference surfaces for a 1-min association followed by a 3-min dissociation. The surfaces were regenerated with a 30 s injection of 3 M MgCl2 after each cycle. Double-referenced titration data was globally fit to a 1:1 Langmuir binding model to determine the association rate constant, ka (M−1 s−1), and the dissociation rate constant, kd (s−1), using the BIAcore T200 Evaluation Software version 2.0 (GE Healthcare). The equilibrium dissociation constant was calculated as KD (M) = kd/ka.

### 2.11. *Liquid Chromatography Mass Spectrometry (LC-MS or Intact Mass)*

Sample was diluted to 0.2 mg/mL with 50 mM ammonium bicarbonate and 4 μL was injected to a POROS R2/10 2.1 × 30 mm Column (Life Technologies, Carlsbad, CA, USA, 1-1112-12). A gradient from 30 to 58% Buffer B (acetonitrile, 0.1% formic acid) in Buffer A (water, 0.1% formic acid) was applied to the chromatographic column with flow rate of 100 μL/min. Data was acquired on a Waters Synapt G2-S Mass Spectrometer and deconvoluted to monoisotopic and singly-charged using the Waters MassEnt 1 software.

### 2.12. *Storage Stability Study*

All samples were formulated in 20 mM sodium acetate pH 5.5 at 5 mg/mL. Solutions were filtered through a Millex-GV 0.2 μM PVDF 33 mm syringe filter (Millipore, Burlington, MA, USA) and aliquoted into 2 mL screw-top microcentrifuge tubes (Fisher). Samples subjected to accelerated storage were placed into a temperature-controlled stability chamber (Thermo Scientific) at 40 °C. Samples were pulled at specific timepoints and then analyzed for purity by SE-UPLC and cSDS.

### 2.13. *Crystallization and Data Collection*

Crystallization plates were set up in 3 sub-well plates (Intelli, Art Robbins) by vapor diffusion using Mosquito (TTP Labtech, Boston, MA, USA) at 4, 18, and 30 °C, and images were acquired using RockImager 1000 (Formulatrix Bedford, MA, USA). Crystals appeared in well of H7 of JCSG Plus Screen (Molecular Dimensions, Maumee, OH, USA) within a few hours and were fully grown after 3 days. Crystals (length, 80–200 μm) were present in condition H7 (0.2 M ammonium sulfate, 0.1 M Bis-Tris pH 5.5, 25% *w/v* PEG 3350) in 1:1, 2:1 and 1:2 protein to precipitant ratio in 200 nL drops. Further optimization of condition resulted in optimal crystal in 2:1 protein to precipitant ratio at 30 °C of 200 nL drops. Crystals were cryo-protected in reservoir solution supplemented with 5% glycerol and flash-cooled in liquid nitrogen. We noticed that crystals harvested after 2–3 days resulted in optimal diffraction. Data collection was performed at the Industrial Macromolecular Crystallography Association (IMCA) beam line, sector 17 of the Advanced Photon Source (APS) at the Argonne National Laboratory (ANL, Lemont, IL, USA). Data were collected at a wavelength of 1.0 Å using a Pilatus 6M detector (Dectris A G, Baden Dättwil, Switzerland). The data were processed using the autoPROC [16,17] automated processing software. AutoPROC utilizes XDS for indexing and integration and, AIMLESS for scaling, POINT LESS for data analysis, and STARANISO for applying anisotropic diffraction limits.

### 2.14. *Structure Determination and Model Building*

The structure was solved by Molecular Replacement using MOLREP [18] and Phaser [19]. The partial model was further extended by AutoBuild [20]. The structure was then refined using autoBUSTER [21] and phenix.refine [22]. The initial maps had poor density for several regions, including some of the CDR-like loops, which were removed from the model and gradually rebuilt during refinement. The electron density map was consistent with most sequence substitutions and insertions or deletions between the starting molecular replacement model and the final structure. The sequence was manually corrected using COOT [23]. The resulting structure was refined using Phenix and rebuilt several times leading to final values of $R_{free}$ and $R_{work}$. The final model contained 3 dimers in the asymmetric unit.

### 2.15. *Constructs Used, Sequence Alignment, and Numbering*

All mAbs, Fabs, HC2s, and Fd2 molecules used in this study were prepared by gene synthesis. Full length mAb versions contained all residues within the $V_H$ and CH domains. mAbs A and B were humanized from mouse and rat immunization campaigns, respectively. For full-length HC2 molecules, the entire HC was used without alterations or truncations. Mutated HC2 molecules were prepared by site-directed mutagenesis. For design of Fabs and Fd2s, engineered IgG1 HC constructs were terminated just prior to the hinge region (or residues 1-224 for Fab-A and Fd2-A and residues 1-235 for Fab-B and Fd2-B) ending in the conserved sequence THT. For the LC, the full-length LC sequence was used (residues 1-219 for Fab-A and residues 1-213 for Fab-B). Both LC sequences terminated in the conserved cysteine involved in the HC:LC intermolecular disulfide. Sequences were aligned and CDRs annotated using proprietary Abacus$^{TM}$ Antibody & Engineering Analysis software using standard pre-sets and reference antibody sets. Default numbering referenced throughout the text is based on sequential numbering, with the exception of the sequence alignments generated by Abacus$^{TM}$, where default sequential numbering was used throughout the alignment by the software.

## 3. Results

### 3.1. Observation and Identification of Expressed and Purified Humanized HC Dimer

Humanized heavy-chain dimer (HC2) formation was observed for two different monoclonal antibodies from transient HC/LC cotransfection followed by Protein A affinity capture. For a recombinant IgG1 mAb, designated here as "mAb-A", a species corresponding to an HC:HC homodimer complex was apparent by NR-cSDS and a lower molecular weight species was partially resolved by SE-UPLC (Figure 1). No other species aside from expected HC and LC were detected by Red-cSDS, although a higher ratio of HC:LC was observed. In Figure 1, representative SE-UPLC and cSDS profiles are shown for mAb-A (Figure 1A,B respectively), along with the verification of covalent full-length HC dimer by intact mass spectrometry (Figure 1C). By NR-cSDS, the full-length IgG1 HC2 species was measured at 4.9% for a standard tube spin 10 mL culture expression (See Table 1 for data). This species is designated "HC2-A". Other constructs corresponding to this same molecule, such as the Fab, Fd portion ($C_H1 + V_H$), or full-length mAb, are designated "Fab-A", "Fd-A", "mAb- A", and so on, deriving from the parent molecule A full-length mAb construct. For other molecules evaluated herein (such as B, C, and D), the same nomenclature applies.

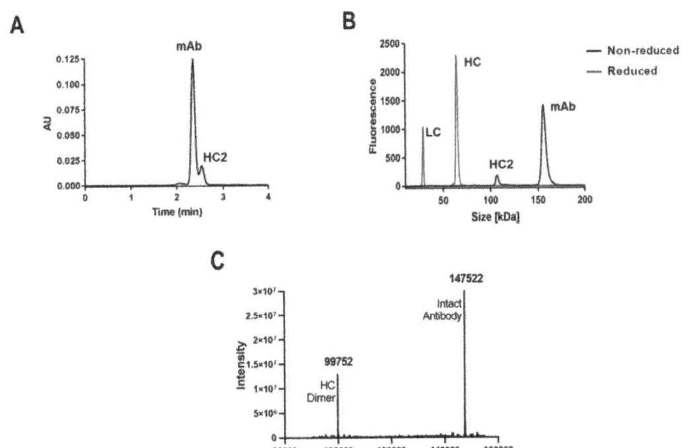


**Figure 1.** Detection of HC2 species in humanized mAb-A molecule by (**A**) SE-UPLC and (**B**) NR- and Red-cSDS (profiles overlaid). (**C**) Confirmation of HC2-A species by intact mass spectrometry from a mixture of mAb and HC2 species.

**Table 1.** Expression variables, codon usage, and isotype evaluated for their effects on mAb-A HC2 formation for 10 mL small-scale transient cultured CHO cells. Yield in milligrams refers to the amount of mAb expressed and affinity captured by ProA except for the HC-only transfection, where yield is the amount of HC2 expressed and purified. Control IgG1 mAb did not yield measurable amounts of HC2 with both HC-only transfection and HC/LC cotransfection.

| Culture Conditions | Yield (mg) | %HC2 by NR-cSDS | Yield (mg) | %HC2 by NR-cSDS | Yield (mg) | %HC2 by NR-cSDS |
|---|---|---|---|---|---|---|
| | IgG4 mAb-A | | IgG1 mAb-A | | Control IgG1 mAb | |
| **HC only Transfection** | 1.0 | 99.7 | 1.3 | 98.0 | 0.0 | 0.0 |
| **Co-transfection with LC** | | | | | | |
| Tube spin control | 2.5 | 7.4 | 2.8 | 4.9 | 1.9 | 0.0 |
| 5x LC chain | 2.1 | 0 | 1.8 | 0 | N/A | |
| 20% coding DNA | 1.9 | 2.8 | 1.6 | 15.7 | N/A | |
| Day 4 harvest | 0.5 | 3.5 | 1.3 | 2.0 | N/A | |
| No enhancer | 2.8 | 2.6 | 2.2 | 2.6 | 0.3 | 0.0 |
| No Temperature shift | 0.9 | 2.3 | 1.0 | 2.7 | 0.5 | 0.0 |
| Shake flask | 2.3 | 5.0 | 2.4 | 2.4 | 2.1 | 0.0 |
| Codon optimization 1 | 3.0 | 3.5 | N/A | | N/A | |
| Codon optimization 2 | 0.1 | 1.6 | N/A | | N/A | |

### 3.2. *Effect of Expression Variables on HC Dimer Expression*

Initially, it was not clear to us what were the root causes of HC dimer formation; specifically, if it was artifactual expression or an inherent property of the antibody sequence driving its assembly. Therefore, we investigated the effects of isotype (human IgG1 versus IgG4), expression variables and DNA codon use on HC2 formation. mAb-A was evaluated as IgG1 and IgG4 (S228P) [24] isotypes alongside a negative control mAb known not to form HC2s. This control is a typical recombinant mAb used as an expression control known for its robust titers and yield. Expression variables such as shaking method, enhancer addition, culture time, and codon optimization were explored for both IgG1 and IgG4 isotypes containing the same variable sequence, and levels of HC2 were quantified by cSDS. In Table 1, these expression conditions are summarized for all molecules across all expression conditions for HC/LC cotransfection and expression. For the controlled "tube spin" condition, both HC-only transfection and HC/LC cotransfection were evaluated. In a 10 mL tube spin transient culture from HC- only transfection, greater than 1 mg of HC2 was captured (or >100 mg/L yield from culture supernatant) for both IgG1 and IgG4 isotypes at a purity of ≥98% by NR-cSDS. Therefore, HC-only expression of mAb-A resulted in efficient expression of HC2 and was affinity captured at high purity. For HC/LC cotransfections, the percent of HC2 quantified by NR-cSDS was approximately 5–7% for both isotypes and ranged from 1.6% to 15.7% across all expression conditions and codon variations, indicating that HC dimer levels are somewhat variable but also robust across experimental conditions. Therefore, it was clear that HC dimer formation was not a consequence of a particular expression or process-related artifact. Our further characterization and investigation of these novel HC2 domains thus utilized the IgG1 isotype arising from standard expression conditions and standard purification.

### 3.3. *Small-Scale Expression of Other Recombinant mAbs and HC Dimer Occurrence*

Using 10 mL small-scale transient tube spin cultures, 12 additional random recombinant mAbs against unique targets were expressed as HC-only transfections and <0.1 mg HC2 or total protein was Pro-A captured for 11 out of 13 (mAbs-C to -M, Table 2). For all the molecules designated C-M that were expressed as HC/LC (mAb) cotransfections, <0.1% potential HC dimer was quantified by

NR- CSDS. However, mAb-B was found to efficiently express HC2 in both HC/LC cotransfections and HC-only transfections. Therefore, as expected, HC dimer formation was unique to certain antibody molecules and sequences and not a common or frequent phenomenon across various antibody sequences. In Table 2, the HC/LC germline families are also tabulated for each mAb. Four of the total 13 sequences are considered rare or uncommon (<10 occurrences out of 358 sequences containing kappa light chains evaluated in Jayaram, et al.), and are highlighted red [25]. Of these four, two express this unique HC dimer at significant levels for both mAb and HC-only expression; specifically, mAbs A and B. The other nine molecules that do not significantly form HC2 all have common or frequently arising germlines (>20 occurrences out of 358 antibody sequences containing kappa light chain evaluated).

**Table 2.** For a panel of 13 recombinant antibodies, tabulated are the HC/LC germlines. These are highlighted red if considered uncommon (<10 occurrences) and unhighlighted if common (>20 occurrences out of 358 kappa-LC containing sequences) [25]. Also tabulated are the % HC dimer (HC2) quantified by NR-cSDS from HC/LC cotransfection (mAb production) and the amount in mg of HC2 species obtained from HC-only small-scale transfection (10 mL total volume).

| mAb | HC/LC Germlines | % HC2 | HC-Only (mg) |
|---|---|---|---|
| A | HV1/KV2D | 4.9 | 1.3 |
| B | HV4/KV1 | 46.2 | 1.5 |
| C | HV3/KV3 | <0.1 | <0.1 |
| D | HV1/KV2D | <0.1 | <0.1 |
| E | HV1/KV1 | <0.1 | <0.1 |
| F | HV3/KV1 | <0.1 | <0.1 |
| G | HV1/KV1 | <0.1 | <0.1 |
| H | HV1/KV1 | <0.1 | <0.1 |
| I | HV3/KV1 | <0.1 | <0.1 |
| J | HV4/KV2 | <0.1 | <0.1 |
| K | HV1/KV1 | <0.1 | <0.1 |
| L | HV3/KV1 | <0.1 | <0.1 |
| M | HV3/LV3 | <0.1 | <0.1 |

Highlights denote infrequent occurance.

### 3.4. *Large-Scale HC-Only Expression of 4 mAbs*

At one liter or greater scale in CHO transient transfection, we expressed multiple full-length IgG1 HC antibody sequences designated as HC2-A to -D (Table 3) and ProA affinity captured the HC2 molecule of interest. Like HC2-A, HC2-B efficiently expresses as an HC dimer and even higher levels of HC2 are observed from HC+LC cotransfection (46.2%, See Table 2). Following HC-only transfection and expression, both HC2-A and HC2-B also were ProA affinity captured to relatively high purity and yield. In Table 3, HC2-A and HC2-B were expressed and ProA captured with a yield of 61 mg/L and 103 mg/mL respectively, from simply an unoptimized transient CHO expression, levels that are quite typical for transiently expressed mAbs. By NR-cSDS, purities were measured at 93.8% and 91.2% respectively for HC2-A and HC2-B, and greater than 96% by Red-cSDS for both. SE- UPLC and cSDS purities were increased to >98% following size-exclusion purification (data not shown) for further characterization, storage stability, and crystallization. Two other molecules (HC2- C and HC2-D) with little propensity to form HC dimer were also evaluated. Recall the mAb forms for these molecules exhibited <0.1% potential HC dimer in small-scale 10 mL HC/LC cotransfection (Table 2). For these two molecules that were HC-only transfected and expressed at large-scale, low amounts of HC dimer were observed at worse purity/heterogeneity, specifically less than 70% by NR-cSDS for HC2-D, indicating that other mAbs may be able to form low levels of lower quality HC2 from HC-only transfection at large-scale, but at insignificant levels during HC/LC cotransfection.

**Table 3.** Yields and SE-UPLC, NR-cSDS, and Red-cSDS purities tabulated for four HC2s expressed as HC-only transient transfections at 3 L scale followed by ProA affinity capture.

| Molecule | Yield (mg/L) | % SE-UPLC Purity | % NR-cSDS Purity | % R-cSDS Purity |
|---|---|---|---|---|
| **HC2-A** | 61 | 88.2 | 93.8 | 97.4 |
| **HC2-B** | 103 | 84.3 | 91.2 | 96.5 |
| **HC2-C** | 5 | 80.6 | 92.6 | 92.2 |
| **HC2-D** | 0.7 | 72.8 | 67.8 | 66.5 |

### *3.5. Accelerated Storage Stability of an HC2 Compared to mAb*

To evaluate whether a full-length HC dimer has good storage stability and potential developability, we evaluated ProA/SEC purified full-length IgG1 mAb-A and HC2-A side-by-side on accelerated one-month storage at 40 °C and surveyed common degradants such as aggregation and fragmentation. Storage stability is a critical aspect of drug product development, and accelerated storage stability at elevated temperatures typically forecasts long-term storage at normal temperatures [26]. In Table 4, the % NR-cSDS purities and % SE-UPLC purities are tabulated at initial, 2 week, and 4 week timepoints for both mAb-A and HC2-A. The % of high molecular weight species quantified by SE-UPLC is also tabulated. NR-cSDS analysis was performed in order to mainly assess fragmentation and SE-UPLC was performed to assess both aggregation (HMW species formation) as well as potential LMW species or fragments. Remarkably, after 4 weeks at 40 °C, HC2-A has better NR-cSDS and SE-UPLC purity attributes versus the full-length mAb-A; by NR-cSDS and SE-UPLC, mAb-A had 96.1% and 98.7% purity, while HC2-A had 98.1% and 99.2%, respectively. Both mAb-A and HC2-A had similar and low levels of SE-UPLC HMW species or aggregation. These findings clearly indicate that HC2 antibodies may have excellent stability comparable to or better than mAbs.

**Table 4.** Formulation conditions were 5 mg/mL for both mAb-A and HC2-A in 20 mM sodium acetate pH 5.5, a representative formulation condition. Both samples were stored up to 4 weeks at 40 °C. Column representations are % purity by NR-cSDS, % purity by SE-UPLC, and % High Molecular Weight species (HMW) quantified by SE-UPLC. Molecules were purified by ProA affinity chromatography followed by SEC polishing.

| Molecule | Timepoint | NR-cSDS | SE-UPLC | % HMW |
|---|---|---|---|---|
| | Initial | 99.1 | 99.6 | 0.4 |
| **mAb-A** | 2 wk 40 °C | 97.8 | 99.2 | 0.4 |
| | 4 wk 40 °C | 96.1 | 98.7 | 0.5 |
| | Initial | 100 | 99.7 | 0.3 |
| **HC2-A** | 2 wk 40 °C | 99.5 | 99.5 | 0.5 |
| | 4 wk 40 °C | 98.1 | 99.2 | 0.8 |

### *3.6. Analysis of Two Fd ($V_H$ + $C_H1$) Domains (Fd2-A and Fd2-B)*

To determine if HC2-A and HC2-B, which were shown to efficiently express full-length HC2 from both transient HC/LC cotransfection and HC-only transfection, can also form HC dimers using only the variable $V_H$ and constant $C_H1$ segment with no hinge and Fc domain present, the "Fd" ($V_H$ + $C_H1$) domain was expressed (designated as dimeric Fd2-A and Fd2-B as follows). These constructs were designed to end just prior to the IgG1 hinge, which contains two intermolecular disulfides. Therefore, the expected Fd2 product should only have one observable intermolecular disulfide bond analogous to Fabs and as previously reported [14]. In Figure 2A, NR- and Red-cSDS profiles are shown for both HC- only Fd2-A and Fd2-B molecules that were $C_H1$ affinity purified. For both molecules, Fd-only expression followed by $C_H1$capture yields >98% purity by NR- and Red-cSDS. High SE-UPLC purities are also observed (>95%) that are easily polished by SEC purification to >99% SEC-UPLC purity like full-length HC2s (data not shown). Further, both molecules clearly form a covalent dimer by NR-cSDS, since a prominent peak at approximately 50 kDa is observed by NR-cSDS, and that this same molecule is reducible by DTT into monomeric Fd domains of approximately 25kDa for both Fd2-A and Fd2-B. The identity of these species was confirmed by intact mass spectrometry (data not shown). This demonstrated that the two different Fd2 species could be assembled, folded, secreted into the culture supernatant, and affinity captured followed by purification of the desired product without the presence of the Fc domain. This also revealed that the $V_H$ and $C_H1$ domains alone were sufficient to form a reducible covalent homodimer, without involvement by the Fc domain, where the two Fd domains are presumed to be linked by a covalent, intermolecular, and reducible disulfide bond. In Figure 2B, representations of the constructs evaluated thus far are shown and include a full-length mAb and HC2 as well as Fab and Fd2 domains. There, the Fd2 domains are shown to be associated biophysically and structurally as is later demonstrated.

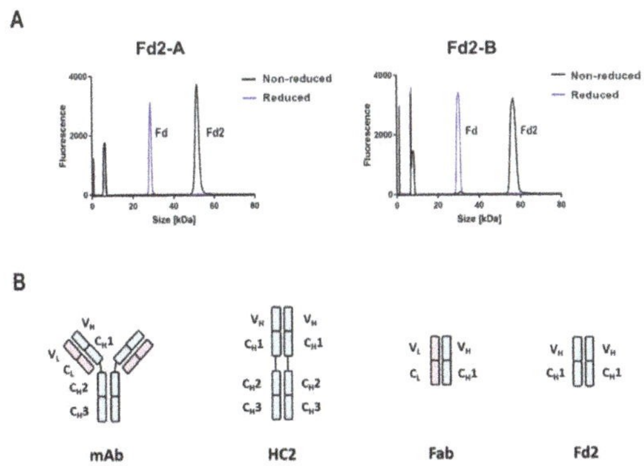


**Figure 2.** (**A**) Overlaid NR-cSDS and Red-cSDS profiles for both Fd2 forms Fd2-A and Fd2-B. The intact Fd2 dimer is reducible into monomeric Fd domains by DTT and appears as a Fd monomer by Red-cSDS. Peaks below approximately 10 KDa are system-related peaks and include the internal MW standard. (**B**) Diagrams representing a mAb, HC2, Fab, and Fd2. The HC is colored light blue and the light chain is light red. Each individual variable and constant domain composing the HC and LC are labeled. Intermolecular disulfides are not shown but exist in the hinge region and between $C_L$ and $C_H1$ as well as between the $C_H1$ domains in the HC-only structures. HC-only Fd2 domains ($V_H$ + $C_H1$) are shown here to be associated.

### *3.7. Differential Scanning Calorimetry (DSC) Analysis of mAb, HC2, Fab, and Fd2 Forms*

Along with full-length mAb and HC2, corresponding Fab and Fd2 truncations were evaluated by Differential Scanning Calorimetry (DSC) to probe differences in thermal stability for both A and B molecular forms. In Figure 3, DSC profiles and tabulated $T_m$ data are shown for mAb, HC2, Fab, and Fd2 forms. For both A and B, mAb and HC2 forms have similarly high, acceptable melting transitions and three in total for each molecule. For mAb-A and HC2-A, $T_m1$ values of 70.4 °C and 65.5 °C were obtained, respectively. For mAb-B and HC2-B, $T_m1$ values of 71.4 °C and 69.5 °C were obtained, respectively. For the Fab and Fd2 constructs, both A and B molecules also had similar and high melting transitions (two transitions for each molecule). Fab-A and Fd2-A $T_m1$ values were 71.5 °C and 75.5 °C, respectively, with the Fd2 (HC-only domain) having higher thermal stability versus the Fab. For the B forms, the Fab and Fd2 $T_m1$ values were 86.3 °C and 78.3 °C, respectively. In the latter case, while the Fd2 construct had lower thermal stability than the Fab counterpart, it still possessed high thermal stability by DSC with a $T_m1$ nearly 80 °C. It is important to note that these Fd2 molecules can be expressed and purified to a fairly homogenous state (See Figure 2). Overall, DSC measurements clearly show that HC2/Fd2 forms for two different antibodies have high thermal stability on par with conventional mAbs/Fabs.

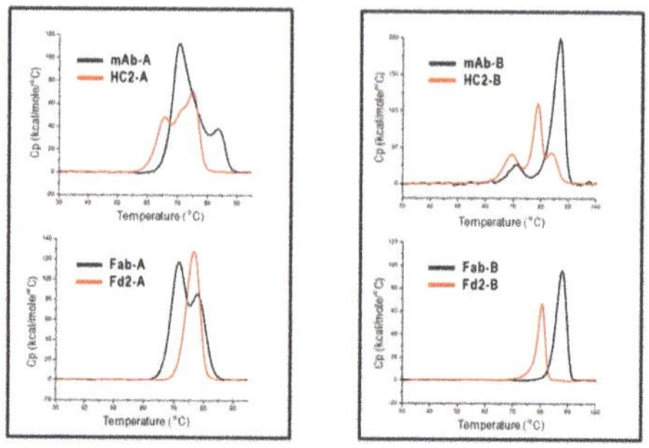


| Species | DSC Tm1 (°C) | DSC Tm2 (°C) | DSC Tm3 (°C) |
|---|---|---|---|
| mAb-A | 70.4 ± 0.03 | 75.4 ± 0.17 | 83.7 ± 0.04 |
| HC2-A | 65.5 ± 0.05 | 71.7 ± 0.10 | 75.5 ± 0.04 |
| Fab-A | 71.5 ± 0.01 | 78.2 ± 0.02 | N/A |
| Fd2-A | 75.5 ± 0.24 | 77.6 ± 0.1 | N/A |
| mAb-B | 71.4 ± 0.17 | 84.4 ± 0.50 | 87.1 ± 0.07 |
| HC2-B | 69.5 ± 0.05 | 78.8 ± 0.02 | 84.4 ± 0.04 |
| Fab-B | 86.3 ± 0.69 | 88.3 ± 0.17 | N/A |
| Fd2-B | 78.4 ± 0.54 | 80.6 ± 0.05 | N/A |

**Figure 3.** DSC profiles and tabulated $T_m$ data for mAb, HC2, Fab, and Fd2 constructs for both A and B molecules. DSC profiles in the upper right box are for the A forms; upper left for the B forms. Below are the tabulated $T_m$ values for each species.

### *3.8. Formation and Analysis of (Non-Covalent) Fd2 Molecules without an Intermolecular Disulfide Bond*

While the Fd domain forms a reducible, covalent dimer (Fd2) for two molecules with different HC framework germlines (HV1 and HV4 respectively) and CDR sequences that bind to two separate biological targets, we wanted to evaluate if these molecules could be expressed, secreted, and dimerized in solution without an intermolecular disulfide bond. Stoyle and coworkers reported that the 5th Cysteine from the N-terminus (Cys219 in Mab-A and designated herein as "5C") is likely responsible for intermolecular disulfide formation between Fd domains [14]. This is the same HC cysteine residue responsible for forming the conserved Fab intermolecular disulfide between the HC and LC [27]. However, it is not known if this is conserved throughout observed HC dimers and further, if the Fd domains can self-dimerize non-covalently in solution. In other words, do Fd2 or HC2 dimers form a true dimerization interface in solution with substantial and relevant interface contacts that is apparent in mAb or Fab forms. When the corresponding 5C-Cys is mutated to serine for both molecules A and B (Cys219Ser in the case of mAb-A, diagrammed in Figure 4A), the Fd domain is still efficiently expressed and purified to high purity. Further, in Figure 4B, denaturing NR-cSDS analysis reveals an Fd monomer of approximately 25 kDa in size for both Fd2-5C-A and Fd2-5C-B, corresponding to the expected molecular weight. Therefore, removal of the disulfide originally responsible for HC-LC disulfide pairing in the Fab domain by "5C" cysteine to serine mutagenesis, also prevents the HC2 (Fd2) from forming a similar intermolecular disulfide bond. In Figure 4B, online SEC-MALS profiles are shown for both Fd2-A and Fd2-B with the 5C disulfide knock-out mutation (Fd2-5C-A and Fd2-5C-B), which elute and are detected from the sizing column at low μM concentration. In solution, a partially resolved mixture of monomeric Fd and non-covalent Fd2 is apparent for Fd2-5C-A. For Fd2-5C-B, non-covalent dimeric Fd2 species are predominant at an average Mw of 49.0 kDa, therefore indicating a more stable non-covalent dimer. We believe these dimeric species do not represent aggregated species, especially since previous accelerated stability work with HC2-A does not result in appreciable aggregation by SE- UPLC even after 1-month storage at 40 °C (See Table 4). Because the monomeric and dimeric species for Fd2-5C-A are not fully resolved, the measured Mw values will be an overestimation for the monomeric form and underestimation for the dimeric form and are tabulated in Figure 5.

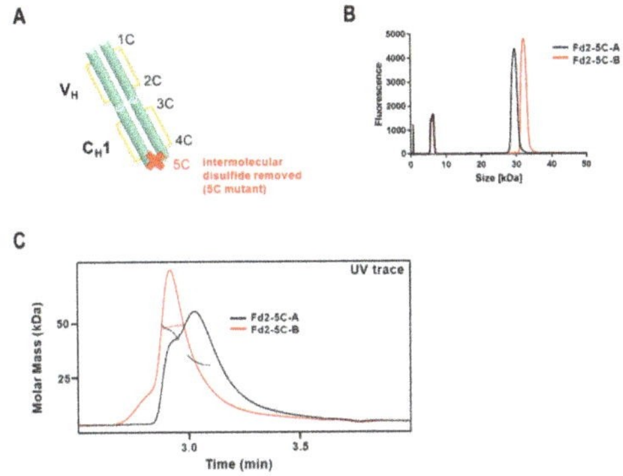


**Figure 4.** Representation of the Fd2-5C dimer variant and experimental data for both Fd2-5C-A and Fd2-5C–B molecules. (**A**) Representation of the dimeric Fd2 domain in cyan, with the $V_H$ and $C_H1$ regions labeled and segmented. Each disulfide bond is represented in yellow, and the location of each cysteine is labeled for one Fd domain (1C-5C counting from the N-terminus). The intermolecular disulfide bond connected by the 5C cysteines is removed by cysteine to serine mutagenesis and thus marked with a red "X". (**B**) Overlaid NR-cSDS electropherograms for both Fd2-5C-A (black trace) and Fd2-5C-B (red trace) and (**C**) Overlaid SEC-MALS profiles for both Fd2-5C-A (black trace) and Fd2-5C-B (red trace). Results are plotted as molar mass (kDa) versus time with each chromatogram generated by UV absorption. Measured molar masses are overlaid on each chromatogram (Fd2-5C-A in black and Fd2-5C-B in red).

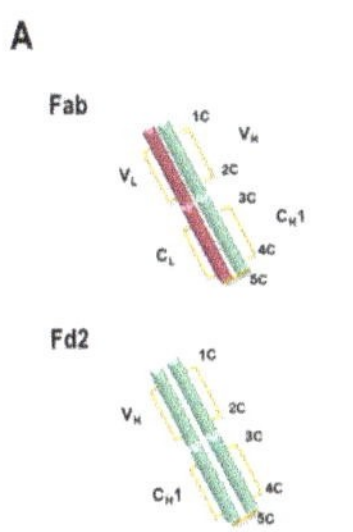


| Chains | Variant | SEC-MALS (kDa) | $T_m1$ (°C) | Secreted species by MS |
|---|---|---|---|---|
| Fd + LC | Fab-A | 48.7 | 71.2 | Fab+Fd2 mixture |
| Fd(C1) + LC | Fab-1C-A | NA | NA | Low titer |
| Fd(C2) + LC | Fab-2C-A | 48.6 | 62.3 | Covalently linked Fab only |
| Fd(C3) + LC | Fab-3C-A | 48.4 | 71.4 | Covalently linked Fab only |
| Fd(C4) + LC | Fab-4C-A | 47.9 | 71.0 | Covalently linked Fab only |
| Fd(C5) + LC | Fab-5C-A | 49.8 | 70.1 | Non-covalently linked Fab (and Fd2); no covalently linked Fd2; free Fd and LC |
| Fd only | Fd2-A | 49.7 | 74.9 | Fd2 only |
| Fd(C1) only | Fd2-1C-A | NA | NA | Low titer |
| Fd(C2) only | Fd2-2C-A | NA | NA | Fd2 only, Low titer |
| Fd(C3) only | Fd2-3C-A | NA | NA | Low titer |
| Fd(C4) only | Fd2-4C-A | NA | NA | Low titer |
| Fd(C5) only | Fd2-5C-A | 46.1 (35.3%); 32.0 (64.7%) | 48.5 | Non-covalently linked Fd2 and free Fd |

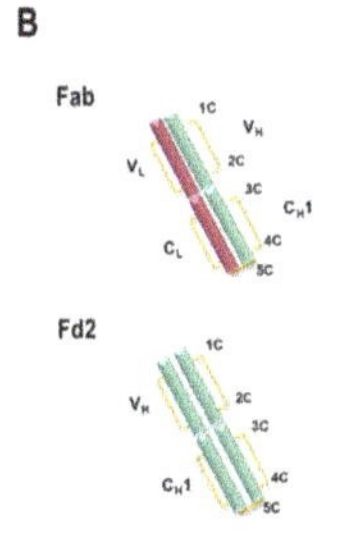


| Chains | Variant | SEC-MALS (kDa) | $T_m1$ (°C) | Secreted species by MS |
|---|---|---|---|---|
| Fd + LC | Fab-B | 50.5 | 78.2 | Fab+Fd2 mixture |
| Fd(C1) + LC | Fab-1C-B | 44.2 | 74.1 | Covalently linked Fab only |
| Fd(C2) + LC | Fab-2C-B | 48.1 | 78.7 | Fab+Fd2 mixture |
| Fd(C3) + LC | Fab-3C-B | 40.9 | 84.5 | Covalently linked Fab only |
| Fd(C4) + LC | Fab-4C-B | 44.5 | 80.8 | Covalently linked Fab only |
| Fd(C5) + LC | Fab-5C-B | 54.9 | 63.0 | Non-covalently linked Fab (and Fd2); no covalently linked Fd2; Free Fd and LC |
| Fd only | Fd2-B | 50.3 | 78.1 | Fd2 only |
| Fd(C1) only | Fd2-1C-B | NA | NA | Low titer |
| Fd(C2) only | Fd2-2C-B | NA | NA | Fd2 only, Low titer |
| Fd(C3) only | Fd2-3C-B | NA | NA | Low titer |
| Fd(C4) only | Fd2-4C-B | NA | NA | Low titer |
| Fd(C5) only | Fd2-5C-B | 49.0 | 63.2 | Non-covalently linked Fd2 and free Fd |

**Figure 5.** Diagrams of the Fab and Fd2 domains along with tabulation of transfected chains, variants, and SEC-MALS Mw and nano-DSF $T_m1$ results for both molecules A (**A**) and B (**B**). The Fab domain is shown as the assembly of the LC ($V_L$ + $C_L$, red) and Fd ($V_H$ + $C_H1$, cyan). The Fd2 domain is also shown as a homodimer of Fd domains (cyan). Like in Figure 4A, disulfide bonds are depicted in yellow and the position of each cysteine is labeled 1C-5C counting from the N-terminus. Under the variant column is specified which cysteine is mutated to serine (e.g., Fab-1C-A has the first cysteine from N-terminus mutated to serine). Species identified by intact mass are listed. "NA" denotes no measurement since insufficient material was expressed and purified for characterization.

### 3.9. Effect of Disulfide Mutagenesis on Fab and Fd2 Formation and Thermal Stability

Subsequently, we wanted to evaluate the effects of all disulfides in the $V_H$ and $C_H1$ regions on Fd2 expression and stability for both A and B molecules. Therefore, each $V_H$ and $C_H1$ cysteine was individually mutated to serine for both Fab and Fd2 constructs on both A and B molecules. Based on known IgG1 disulfide architecture [27], Cysteines 1-4 (for mAb-A, sequential numbering is Cys22 corresponds to "1C", Cys96 is "2C", Cys143 is "3C", Cys199 is "4C") form a pair of conserved intramolecular disulfides within the Fd portion of the Fab domain and pair as 1C-2C and 3C-4C respectively. The 5th cysteine from the HC N-terminus (Cys219 for mAb-A and Cys230 for mAb-B, or "5C") forms the intermolecular disulfide between the Fd and LC of the Fab domains as aforementioned. These molecules will be designated as the construct format, followed by location of cysteine mutagenesis, followed by A or B molecule (e.g., "Fd2-2C-A"). In Figure 5A,B, SEC-MALS and nano-DSF results, along with species observed, are tabulated for each of the resulting variants. Nano-DSF was utilized here to evaluate $T_m$ in a more high-throughput manner for many variants and generally yields results comparable to DSC. In WT Fab expression of both A and B molecules, a mixture of Fab and Fd2 species are observed. Since both Fab and Fd2 are ~50Kda in mass, SEC-MALS measured the Mw to be ~50KDa for this mixture for both A and B. For both Fab-A and Fab-B, Fab expression occurred for nearly all cysteine mutants with the exception of low Fab-2C-A expression. For Fab 1C-2C mutants, similar or lower $T_m1$ values are observed by nano-DSF analysis, while similar or higher $T_m1$ values are observed for 3C-4C mutants. SEC-MALS results for WT Fab and each of the 1C-4C Fab variants indicate that an intact Fab is expressed and formed as expected, with a measured Mw of ~50 KDa. When the intermolecular Fab disulfide is broken through 5C mutagenesis, $T_m1$ is similar for Fab-5C-A but reduced ~15 °C for Fab-5C-B. These Fab-5C species are intact, non-covalent Fabs indicated by SEC-MALS and NR-cSDS results (intact Fab by native SEC-MALS and composed of monomeric Fd and LC components by denaturing NR-cSDS). When the same mutagenesis approach is applied to HC or Fd-only expressed molecules A and B, similar traits are observed with one notable exception: when any cysteine responsible for intramolecular disulfide formation is mutated in HC-only Fd2 expression, little or no protein expression is observed, and therefore no analysis was conducted. Only for the case of Fd2-2C-A expression was Fd2 even detectable by NR- cSDS. This explains why the same variants also resulted in pure Fab species for the Fab-A and Fab-B 1C-4C mutants, since 1C-4C mutated Fd2 was unable to be expressed to appreciable levels. Similar to the case of the Fabs, when the intermolecular disulfide bond is disrupted by 5C mutagenesis, intact, non-covalently dimerized Fd2 is expressed and purified as revealed by SEC- MALS and NR-cSDS (See also Figure 4). For both Fd2-5C-A and Fd2-5C-B, $T_m1$ is significantly reduced compared to Fd2-A and Fd2-B, respectively. For Fd2-5C-A, $T_m1$ is reduced 26 °C, and for Fd2-5C-B, $T_m1$ is reduced 15 °C. This is similar to the case of the Fab-5C-B molecule, where $T_m1$ is reduced 15 °C when the Fab domain is no longer held together non-covalently.

### 3.10. Crystallization and Structural Evaluation of Fd2-A

To understand the structural basis of HC dimer formation and the nature of HC dimerization, crystallization studies were conducted for the Fd2 domain of molecule A (Fd2-A). Crystals obtained from a sparse matrix screen diffracted to 2.9 Å and the structure was solved by Molecular Replacement (See Table 5 for crystallographic statistics). The resulting model was refined and built several times and led to final values of $R_{free}$ and $R_{work}$ of 26.8% and 19.8%, respectively. The final model consists of six monomers of 215 residues each that form three homodimers of HC:HC (Figure 6A) in the asymmetric unit. This is deposited in the PDB bank as PDB I.D. 7KQY. In the crystal structure, the three molecules within the asymmetric unit are structurally similar but have distinct and different domain orientations. The final electron-density map allowed the positioning of most residues with confidence. In Figure 6B, the Fd2-A structure is shown in ribbon diagram and reveals a highly symmetrical homodimer. The CDR loops are also highlighted and face outward, similar to how CDR loops are positioned in Fabs/mAbs. However, we found these HC2s for molecules A and B to not bind their original targets as expected using a BIAcore SPR affinity assay (data not shown). The electron density corresponding to residues

HC-216 to 224 is very weak, indicating that the loop is disordered. This disordered loop also includes Cys219, which is engaged in an intermolecular disulfide bond to itself in the opposing monomer based on previous studies of HC dimers [14] and our disulfide mutagenesis experiments. While this loop is disordered and therefore not shown, it appears to be in close proximity to the same loop in the opposing Fd monomer, thus allowing for intermolecular disulfide formation. In Figure 6C, the Fd2-A homodimer is shown with all observed intramolecular disulfides shown in red surface. These pairs of intramolecular disulfides are conserved with respect to all IgG1 mAbs. In Figure 6D, the novel Fd2-A complex is superposed with a Fab whose crystal structure was solved and previously reported (PDB ID 5vsi). Specifically, the $C_H1$ and $V_H$ domains were superposed and overlay closely, with a low overall C$\alpha$ RMSD of 1.10, highlighting how similar the HC2 (or Fd2) and Fab structures are. In Figure 6E, the $C_H1$ domains of the Fd2-A and Fab (5vsi) are superposed. Interfacial Fd2-A dimer contacts that differ from 5vsi are shown in red, which includes symmetrical interactions involving the loop containing residues HC-126 through HC-133. Also in Figure 6E and overlaid with Fd2-A, the corresponding Fab loop (HC-131 to HC-138) is highlighted in magenta and is involved in separate interactions with the LC; it is in this region where a different set of unique interactions exist when comparing the Fd2-A and Fab structures (See Supplemental Figure S1 for sequence alignment).

**Table 5.** Crystal data collection and refinement statistics. Values in parenthesis describe the highest resolution shell. Ramachandran refers to backbone dihedral angles.

| *Data Collection* | **HC2-A (7KQY)** |
|---|---|
| Space group | C121 |
| *Unit cell dimensions* | |
| $a, b, c$ (Å) | 169.66, 73.42, 140.41 |
| $\alpha, \beta, \gamma$ (°) | $\alpha = 90$, $\beta = 125.71$, $\gamma = 90$ |
| Resolution (Å)[a] | 114–2.91 (3.02–2.91) |
| Unique reflections | 22,757 (131) |
| Completeness (%) | 73.43 (4.28) |
| Multiplicity | 3.4 (1.7) |
| Wilson B-factors | 54.5 |
| $I/\sigma I$ | 4.85 (1.18) |
| *R*merge | 0.103 (0.471) |
| Rmeas | 0.146 (0.665) |
| *R-pim* | 0.103 (0.471) |
| CC1/2 | 0.979 (0.619) |
| CC * | 0.995 (0.874) |
| ***Refinement*** | |
| Resolution (Å) | 114–2.91 (3.02–2.91) |
| Unique reflections | 22,743 (131) |
| *R*work/*R*free (%) | 19.8/26.8 |
| CC work/CC free | 0.933/0.905 |
| Atoms | 9247 |
| Protein | 9247 |
| Protein residues | 1243 |
| RMSD Bond lengths (Å) | 0.01 |
| RMSD Bond angles (°) | 1.28 |
| Ramachandran preferred/allowed (%) | 88.2/8.7 |
| Clashscore | 10.84 |
| Average B-factor | 41.4 |

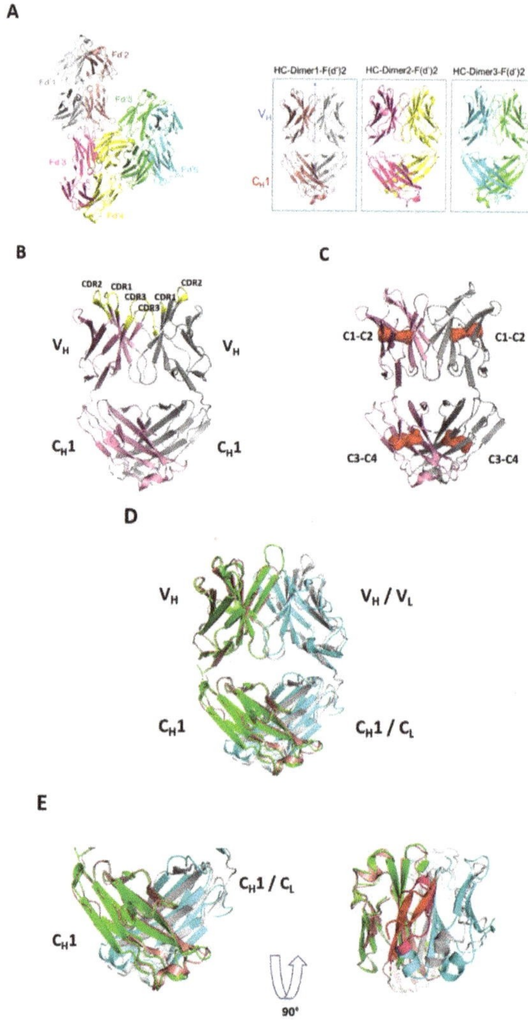


**Figure 6.** (**A**) Three HC or Fd dimers were found in asymmetric unit. All 3 dimers are similar and have the same topology except for minor differences in electron density in flexible regions. "HC- Dimer1-Fd2" (dimer of Fd′1 and Fd′2 domains) was chosen for evaluating structural features such as interfacial contacts, CDR topology, and comparisons to a representative Fab domain. (**B**) Ribbon diagram of the highly symmetrical HC dimer Fd2-A (PDB accession code 7KQY). One Fd chain is colored light purple and the opposing chain is gray. CDR loops are yellow and labeled. This representation is the same HC dimer as "HC-dimer1-Fd2" shown in part A. (**C**) Fd2-A with each intramolecular disulfide bond highlighted red in transparent surface and labeled with the same notation used in Figure 5 (C1, C2, etc., where C1 is disulfide bonded to C2 and C3 bonded to C4) (**D**) Superposition of Fd2-A dimer (shown as salmon for chain A and gray for chain B) and HC-LC of a Fab (PDB accession code 5vsi). The Fab HC (or Fd) chain is in green and the LC is in turquoise. (**E**) Superposition of the Fd2 $C_H1$ domains with 5vsi Fab $C_H1$ and $C_L$ domains (left) and same domain rotated back 90° (right). In the HC2-A dimerization interface, residues from $C_H1$ shown in red are residues HC-126 to HC-133 involved in a symmetrical interaction that is unique relative to the Fab interface. In magenta, this portion of the Fab HC-LC interface (residues HC-131 to HC-138 for 5vsi) are interactions not present in the HC2-A interface.

In the highly symmetrical Fd homodimerization interface, several key and conserved residues are apparent and hold the dimerization interface together (Figure 7A). Involved are residues spanning the $V_H$ (Figure 7A) and $C_H1$ (Figure 7B) domains and include both highly conserved FW and variable residues. In the FD2-A structure, the opposing $C_H1$ domains are in direct contact with each other and the $V_H1$ domains are in direct contact as well, analogous to the Fab structure. Between chains A and B (opposing monomer units within dimer) within the $C_H1$ regions, key intermolecular contacts include side chain contacts between A-His-167 to B-Gln-174 (3.4 Å interatomic distance), side chain contacts between A-Gln-174 to B-His-167 (3.4 Å), backbone contacts between A-Ala-128 and B- Ala-128 (2.8 Å), side chain contacts between A-Pro-126 to both B-Ser-130 and B-Ser-133 (3.5 Å and 3.1 Å respectively), and side chain contacts between A-Ser130 to B-Gln-215 (2.9 Å). Additionally, a symmetrical network of interactions exists between conserved $V_H$ FW residues Tyr-95 and Gln-39 of both chains A and B, where each side chain is contacted at an interatomic distance of 3.2 Å, with Gln- 43 flanking and stabilizing this network of interactions. Gln-39 in both chains also have side chain contacts with each other at a distance of 3.4 Å. Because these residues are conserved, the decision was not to mutate these residues and evaluate their effect on HC dimer formation as they are conserved across all mAbs. In the variable region of Fd2-A, key, highly symmetrical contacts exist mainly between the HC-CDR2 and HC-CDR3 regions (See Figure 7A). Hydrogens bonds are apparent between the side chains of residues HC-CDR2-A-Asp-59 and HC-CDR3-B-Arg-102 (3.4 Å distance), as well as HC-CDR3-A-Arg-102 and HC-CDR2-B-Asp-59 (3.4 Å distance).

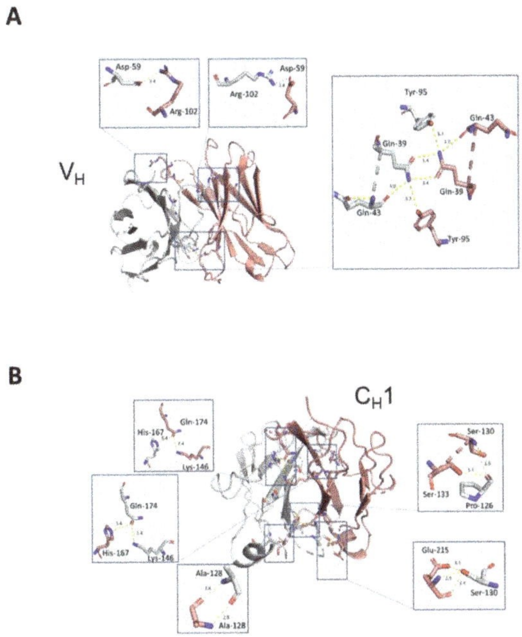


**Figure 7.** (**A**) Key interfacial contacts highlighted in the variable region ($V_H$). (**B**) Key interfacial contacts highlighted in the constant region ($C_H1$). Fd chains A and B are colored salmon and gray.

Many of the apparently conserved interfacial homodimer contacts present in the Fd2-A structure have similar or analogous contacts present in the Fab (HC-LC) interface [28]. In 5vsi, intermolecular side chain contacts between the $V_H$-$V_L$ FWs are observed between HC-Gln-39 and LC-Gln-37, which appear to be substituted by a symmetrical HC-Gln-39 interaction in the Fd2-A structure. Additionally, the Fab HC-Tyr-95 to LC-Gln-37 interaction is replaced by duplicate, symmetrical HC- Tyr-95 to HC-Gln-39 interactions in the Fd2-A structure. In the Fab $C_H1$-$C_L$ interface, intermolecular side chain contacts are observed between HC-His-172 and LC-Asp-137. This appears to be replaced in the Fd2-A structure by a duplicate set of symmetrical HC-His-167 to HC-Gln-174 side chain interactions, where Fd2-A-His-167 aligns with His-172 in the 5vsi Fab. Towards the base of the Fd2-A complex, there is some divergence in contacts and structural alignment with 5vsi Fab, although there are additional contacts with residues in common to both Fd2-A and 5vsi Fab. For instance, in the Fab, an interfacial side chain contact exists between HC-Ser-138 (sequence aligns to Fd2-A-Ser-133) and LC-Phe-177.

### 3.11. Mutagenesis of Fd2-A $V_H$ Interfacial Residues and Their Effect on HC2 Formation

Since the pair of aforementioned Fd2-A R102-D59 interactions are symmetrical between both chains in the homodimer and involve residues unique to that molecule, these residues were mutated to examine their effects on HC2 dimer formation. If these interfacial residues were mutated to different residues that do not favor HC dimer formation, the result should be a lower propensity to form HC dimer. These mutations either removed the side chain or were mutated to other residues present in other mAbs with little tendency to form HC dimer, specifically residues found in mAbs C and D (See Figure 8 for alignment). In Table 6, these results are summarized for % HC dimer formation for HC/LC cotransfection to produce mAbs in CHO transient cells. For mAb-A, HC-CDR2- D59N/A and HC-CDR3-R102A/Q mutations were engineered. For mAb-A, all of these mutations reduced the % HC dimer expressed and purified by several fold (in the range of 7 to 16.3 fold reduction). Further, using a sequence alignment of the $V_H$ domains of both molecules A and B (Figure 8), the residues HC-S58 and HC-Y133 in mAb-B aligned with HC-D59 and HC-R102 in mAb- A, respectively. When similar mutations were made to these residues in mAb-B, namely HC- S58A/Y and HC-Y113A, the % HC dimer expressed and purified was reduced in the range of 6.9 to 27.5 fold. Remarkably, all these mAb variants maintained target binding in the sub-nM range similar to WT using a monovalent BIAcore SPR affinity assay against their respective target antigens (Table 6). Therefore, the Fd2-A crystal structure guided mutagenesis of both mAb-A and mAb-B, which in each case resulted in decreased HC dimer formation propensity. From a developability context, this indicates HC2 formation can be mitigated by site-directed mutagenesis of these sites while not impacting activity; from an engineering perspective to create HC2 molecules, similar reverse mutagenesis may be successful at similar or distinct sites.

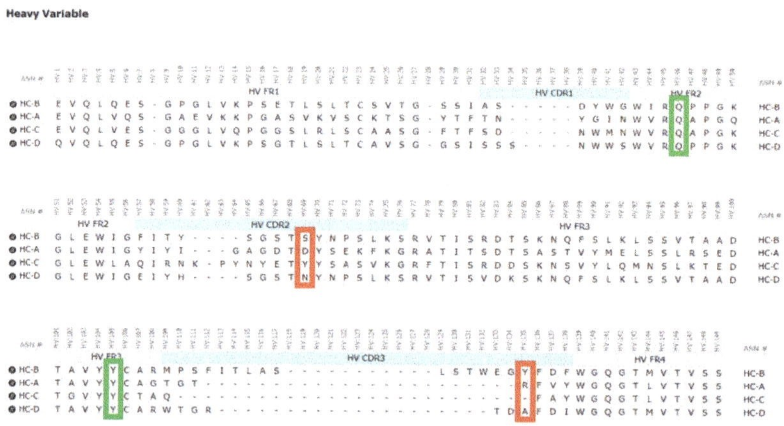

**Figure 8.** Sequence alignment of the $V_H$ domain for the 4 molecules expressed at large scale, or HC- A, -B, -C and -D. Key conserved, interfacial HC2 $V_H$ interactions are outlined in red/green. Variable residues where mutagenesis was attempted are indicated with a red box and involve aligned residues in the HC-CDR2 and HC-CDR3. Due to differences in HC-CDR3 lengths which are common, HC2-A R102 was aligned with HC2-C Q99, or the first HC-CDR3 residue. More conserved interfacial contacts where mutagenesis was not attempted are indicated in a green box.

**Table 6.** For mAb HC/LC cotransfection, effect of HC2 formation for both mAb-A and mAb-B is evaluated for each mutant predicted to reduce HC dimer propensity by the Fd2-A crystal structure. % mAb form and % HC2 quantified by NR-cSDS are tabulated along with fold reduction compared to the wild-type version. Lastly, BIAcore affinities of the mAb (KD) towards recombinant target antigen as the analyte are tabulated.

| Reagent Name | % HC2 | HC2 Fold Reduction | % mAb | KD (nM) |
|---|---|---|---|---|
| **mAb-A WT** | **4.9** | - | **92.5** | **0.29** |
| mAb-A R102Q | 0.5 | 9.8 | **97.7** | 0.3 |
| mAb-A R102A | 0.5 | 9.8 | **98.0** | 0.46 |
| mAb-A D59N | 0.3 | 16.3 | **98.1** | 0.67 |
| mAb-A D59A | 0.7 | 7 | **97.7** | 0.56 |
| **mAb-B WT** | **46.2** | - | **47.4** | **0.11** |
| mAb-B Y113A | 1.7 | 27.5 | **90.7** | 0.41 |
| mAb-B S58Y | 6.7 | 6.9 | **87.9** | 0.27 |
| mAb-B S58N | 6.6 | 7 | **88.0** | 0.23 |

## 4. Discussion

Herein, we structurally and biophysically characterized unexpectedly occurring human heavy-chain only antibodies arising directly from transient CHO expression. We found these non- artefactual HC-only antibodies to be homodimeric in nature, with an extended interface joining both $C_H1$ and $V_H$ domains, and to be linked by a conserved intermolecular disulfide bond similar to how Fabs are linked. Overall, these HC dimers are efficiently expressed with or without the LC and conserved Fc domain. Further, these HC-only molecules have similar or better accelerated storage stability characteristics and thermal stability compared to their mAb or Fab counterparts. Further, the HC dimer is structurally analogous to the Fab, and this is an interesting result that has implications concerning its mechanism of assembly and its potential application as a novel antibody format. The solved crystal structure and structure-guided, rational mutagenesis to the dimer interface reveals for the first time structural and sequence determinants to human heavy-chain only antibody formation.

For both mAbs A and B presented in this study, which are two humanized and functional recombinant antibodies with different germlines (HV1/KV2D and HV4/KV1 respectively) against different and unrelated biological targets, HC-only expression yields stable HC dimers (HC2s). These HC2s are reducible into monomeric components as demonstrated by denaturing R-cSDS of Fd2 forms and NR-cSDS of C219S-Fd2 forms (lacking the intermolecular disulfide, see Figures 2 and 4). In Table 1, we see that expression variables and isotype do not eliminate or significantly impact HC dimer formation for molecule A (HC2-A), and HC dimer remains robust, although varied, across conditions. The only exception to this is with the addition of 5X transfected LC, which can effectively eliminate HC2-A formation but was not considered to be a viable downstream process solution.

Specifically, the HC cysteine involved in the intermolecular disulfide bond connecting each Fd domain in the Fd2 molecule is the same HC cysteine that disulfide links the Fd and LC together in Fabs. Interestingly, when the Fd-Fd (no Fc and hinge disulfides present) intermolecular disulfide bond is broken through C219S/C230S mutagenesis for Fd-A and FD-B respectively, these HC (Fd) dimers are still able to be expressed and yield molecules that are partially or predominantly non- covalently dimerized in solution at low μM concentrations. This was demonstrated by native SEC-MALS, suggesting there is a real dimerization interface with low μM affinity. This result also justified crystallographic studies in an attempt to structurally reveal the nature of this dimerization interface. This same experiment was conducted for Fab expression (HC or Fd +LC), however, since a mixture of non-covalent Fab and Fd2 was formed, it is difficult to parse out comparisons between non-covalent Fd2 and Fab properties.

Interestingly, while both IgG1 mAbs A and B efficiently yield HC dimer both in the full-length and Fd2 (without Fc) forms from HC-only CHO transient transfection, in HC/LC cotransfected CHO transient expression, HC2-B clearly is more efficiently expressed than HC2-A (46.2% vs. 4.9% in control tubespin cultures). While we deemed both of these mAb molecules to have huge developability risks and concerns due to robust HC dimer formation, mAb-B HC/LC transfection almost equally expressed HC2-B versus mAb-B. By SEC-MALS analysis, C230S-Fd2-B was also predominantly a non-covalently intact Fd2 dimer, whereas C219S-Fd2-A was predominately monomeric with some non-covalent dimer present. Considering the amount injected (20 μg) on the SE-UPLC, the measured UV signal, and the volume over which the C219/230S-Fd2-A/B analytes eluted (0.25–0.5 mL), these dimeric Fd2 species were detected by SEC-MALS in the 10–20 μM range. This same C230S-Fd2-B species had a higher $T_m1$ (63 °C) than did C219S-Fd2-A (49 °C), although this difference may have been due to the instability of the predominant monomer form for C219S-Fd2-A at low μM concentrations. Empirically, it is evident that mAb-B had significantly higher levels of expressed HC2 in the presence of LC than did mAb-A and it also qualitatively has a stronger dimer interface in the solution state as suggested by SEC-MALS data. The stronger dimer interface present in Fd2-B versus Fd2-A may explain why Fd2-B forms more robustly in the presence of LC, where apparently HC2 or Fd2 formation is seemingly in direct competition with mAb and Fab formation and is more favored if the Fd2 interface is strengthened. Conversely, when mutations were made to the dimer interface elucidated by the novel Fd2 crystal structure that were predicted to weaken the dimer interface for both HC2-A and HC2-B, HC dimer levels decreased in the presence of LC expression. Therefore, these dimer interface mutants selectively decreased HC dimer formation versus competing HC-LC formation as originally hypothesized.

Conversely, if HC-LC pairing were strengthened or weakened, this should lower or increase HC dimer formation, respectively. When we express mAb-A in the presence of 5X LC, mAb expression dominates and we do not obtain any measurable amounts of HC2-A. This is a clear indication that mAb assembly outcompetes potential HC2 assembly when LC is expressed in vast excess. Moreover, when we evaluate HC/LC germline pairing frequencies, we see that uncommon germlines may increase the likelihood of HC2 formation (Table 2). In a publication by Jayaram and coworkers, 358 HC/kappa-LC human antibody pairings were evaluated, and specific germline frequencies were tabulated [25]. Their findings suggest that germline pairings are not random and that common germline pairings may be attributable to increased stability of the $V_H/V_L$ interface. In mAb-A, the HC and LC germlines are the result of a more uncommon germline pairing, HV1/KV2D, where only four instances out of 358 occurred. However, this uncommon pairing may not be statistically significant when considering that KV2D itself is infrequently encountered. Likewise, mAb-B also has an uncommon germline pairing, HV4/KV1 (nine out of 358 instances). When 11 additional human/humanized recombinant IgG1 and IgG4 mAbs against different targets were evaluated, nine had more common pairings (>20 instances each) and none formed HC dimers appreciably. Although our panel of molecules is limited and anecdotal, our findings may indicate germline pairing preferences play a role in enabling HC dimer formation and also explain their uncommon occurrence. However, clearly there are other intrinsic factors to HC dimer formation which are presumably sequence/structurally dependent. The heavy-chain domain still needs to be formed, folded, and result in a reasonably stable product independent of the LC, a characteristic unique to a certain subset of human HC sequences.

Another example or factor in HC2 formation may include the length of the HC-CDR3, where HC2-B has an unusually long HC-CDR3 of 19 residues (See Figure 8). HC-CDR3 length and composition can have a huge impact on the developability properties of a mAb [29]. Previously, HC-CDR3 length has correlated with negative developability attributes such as aggregation and high viscosity, where 137 clinical-stage antibody therapeutics had a median HC-CDR3 length of 12 residues [30]. For HC/LC transfection of mAb-B, HC2 formation (46.2%) was nearly equal to normal mAb assembly (See Table 2). Stoyle and coworkers reported the number of basic residues to be a determinant of HC2 expression, where an increase in basicity of the HC-CDR3 increased HC2 propensity [14]. However, no basic residues are observed in the HC-CDR3 for HC2-B. In the case of HC2-A, a relatively short HC-CDR3 of seven residues exists, with only one basic residue present. Clearly, HC2 formation may depend on other complex and unknown antibody sequence and structural factors.

It is well known that the molecular chaperone protein, BiP, binds non-covalently to the HC, but not to the HC associated with LC [31]. During mAb assembly, HC-LC pairing and folding in the Fab arm occurs following BiP release from unfolded HC ($C_H1$ + $V_H$ or Fd) as a result of LC assembly [32]. Here in the case of molecules A and B, HC dimer formation occurs with or without the presence of LC, implying that perhaps one Fd domain mimics the LC, and can promote folding of another Fd domain presumably associated to BiP. Remarkably, as depicted in Figure 6D, the novel HC2 or Fd2 complex reported herein overlays nicely with a representative Fab structure with a low overall Cα RMSD of 1.10 Å, revealing that the opposing HC, or Fd, likely mimics the LC structurally during assembly and promotes complete folding and assembly of HC2s/Fd2s. While it is known that HC dimers are common degradant products of formulated mAb (HC2-LC2) drug products [33], resulting from the loss of two LC domains, these HC2s described herein form during expression without LC, and are secreted and captured as fully folded and stable products. Indeed, we show that these novel HC-only antibodies are highly stable with $T_m$s similar or better than their mAb or Fab versions (see Figure 3), where specifically Fd2-A had a measured $T_m1$ of 75.5 °C compared to 71.5 °C for Fab-A. Additionally, we found that full-length HC2-A had excellent accelerated storage stability attributes that were superior to mAb-A, where after 1-month storage at 40 °C, purity attributes by SE-UPLC and NR-cSDS were both >98% (See Table 4).

It is also noteworthy that intramolecular disulfide formation is necessary for HC2 or Fd2 expression. When one intramolecular disulfide bond is removed in the Fab-A and Fab-B forms, LC assembly followed by BiP removal is still able to occur, and Fab folding and secretion is achieved. The result for Fab-A and Fab-B are expressed Fab molecules with suitably high $T_m$ values. This is not unexpected since reduced disulfides can occur naturally in antibodies [34], although incomplete disulfide formation may result in increased aggregation propensity or decreased bioactivity [35–37]. In HC2/Fd2 formation, one Fd domain appears to mimic or replace the role of the LC, however, when one intramolecular disulfide is removed on both Fd chains via mutagenesis, Fd2 assembly is not efficient and little to no product is yielded. This demonstrates that complete disulfide formation is necessary in one or both Fd domains to assemble to the Fd-BiP intermediate(s) and complete Fd2 folding and BiP release. While mutagenesis studies of $C_H1$ cysteines involved in both inter- and intramolecular disulfide formation for a mouse IgG2 enabled expression and secretion of HC-only transcripts [38], we find complete intramolecular disulfide formation a requirement for the expression of two different humanized HC-Abs evaluated in this study.

Structurally, the solved crystal structure of Fd2-A is remarkably symmetrical and involves many conserved residues in the homodimer interface that have similar or analogous interactions in the Fab HC-LC interface. Like Fabs, the CDR loops are facing outward and primed for and amenable to antigen binding. As aforementioned, our novel HC-only Fd2 complex superposes remarkably with a representative Fab structure (5vsi, See Figure 6D), demonstrating that these molecules are quite similar to Fabs, and therefore have potential therapeutic utility. Although HC dimers A and B characterized in this study are not active against the same target antigens that the mAb/Fabs are (both in the sub-nM affinity range), this is fully expected since the entire LC is missing and further, the arrangement and conformation of the HC-CDR loops are very likely to differ between the HC2 and mAb forms. In the typical Fab structure, principal contacts exist between the HC-CDR3 and LC- CDR1. In this HC2 structure for molecule A, the HC-CDR3 is also primarily involved in interfacial contacts, with key contacts existing between the opposing HC-CDR2 chain, which apparently substitutes for the LC-CDR1 in mAbs. The HC2 homodimer structure is also highly symmetrical in nature, which may offer interesting binding and therapeutic properties, and may perhaps be amenable to engineering hetero-HC dimers (composed from two different HCs) that have functional utility. To accomplish this, HC-only domain libraries could be selected against a given target by yeast surface display, as many non-conventional antibodies and scaffold proteins have been engineered using this platform technology [39]. Further, hetero-human-HC dimers may be engineered to associate orthogonally by utilizing interfacial knob-hole interactions, for instance. These approaches may facilitate multi-specific formats in the presence of conventional Fab domains.

Mutations designed to disrupt the Fd2 dimer interface were based on direct $V_H$ contacts revealed by the crystal structure of Fd2-A. These mutations were made for mAb-A and to aligned residues in mAb-B. When these key, interfacial CDR residues present in Fd2-A are mutated to smaller residues or residues present in other non-HC dimer forming mAbs (C and D), HC dimer expression is significantly reduced for both A and B (See Table 6). Likewise, the HC2-A and HC2-B molecules have many other similarities, such as both being linked by an intermolecular disulfide bond and having high thermal stability, indicating that the structures of HC2-A and HC2-B are likely similar and conserved. Additionally, as demonstrated by mutagenesis of interfacial dimer contacts, further engineering is achievable to modulate HC dimer formation through engineering or display technologies. Surely, other sequence or specificity determinants of HC2 formation likely exist and understanding these will further advance potential engineering. The use of engineered and/or antibody generated human heavy-chain dimers can have advantages such as being smaller in nature versus full-length antibodies as well as avoiding LC mispairing in multi-specific formats [40]. We show herein these human HC dimers are highly stable, structurally well-ordered, crystallizable, and their likely conserved and modular nature presents a path forward towards potential engineering and development of these molecules into therapeutic modalities.

## 5. Conclusions

The novel characterization and structural elucidation of dimeric and highly symmetric human heavy chain-only dimers (HC2s) were revealed for the first time and have significant implications to both the mechanism of antibody assembly and their potential application as a novel and functional antibody format. Unlike mAbs or Fabs, these HC2 species are expressed efficiently, independent of the LC, where one HC domain seemingly mimics the LC during assembly. These novel HC2s structurally overlay closely with Fabs, while sharing the same Fab disulfide bond network, secondary folds, and residues involved in their respective HC2 dimer or Fab interfaces. These HC2 species can also be purified to high homogeneity and are highly stable with thermal and acceleration storage stability similar or greater than their mAb or Fab counterparts. We verified residues in both the HC-CDR2 and HC-CDR3 in two different monoclonal antibodies against different targets to dramatically affect HC dimer propensity, highlighting for the first-time distinct sequence specificity determinants for this novel modality. We also elucidated important developability considerations and how to avoid HC2 formation during mAb expression, both from a sequence evaluation and process standpoint. Like Fabs, these molecules also present the CDRs outward for potential antigen binding, which can be exploited to find binding partners and potentially have utility as a novel antibody format.

**Supplementary Materials:** The following are available online at http://www.mdpi.com/2073-4468/9/4/66/s1, Figure S1: HC2-A and HC-Fab Sequence alignment.

**Author Contributions:** Conceptualization, Investigation, and writing, C.M.; Crystallization and Structural Determination, S.B.; Characterization, Y.Y.; Protein expression, J.B.; Protein Engineering, G.R.; Mass Spectrometry, D.T.; SPR Binding, M.H.; Characterization, M.M.; Review and editing, C.S.; Review and editing, L.F.-D. All authors have read and agreed to the published version of the manuscript.

**Funding:** This research received no external funding.

**Conflicts of Interest:** The authors, all employees of Merck & Co., Inc., declare no conflict of interest.

## References

1. Moghimi, S.M.; Rahbarizadeh, F.; Ahmadvand, D.; Parhamifar, L. Heavy Chain Only Antibodies: A New Paradigm in Personalized HER2+ Breast Cancer Therapy. *Bioimpacts* **2013**, *3*, 1–4.
2. Hamers-Casterman, C.; Atarhouch, T.; Muyldermans, S.; Robinson, G.; Hamers, C.; Songa, E.B.; Bendahman, N.; Hamers, R. Naturally occurring antibodies devoid of light chains. *Nature* **1993**, *363*, 446–448.

3. Arbabi-Ghahroudi, M. Camelid Single-Domain Antibodies: Historical Perspective and Future Outlook. *Front. Immunol.* **2017**, *8*, 1589.
4. Flajnik, M.F.; Deschacht, N.; Muyldermans, S. A case of convergence: Why did a simple alternative to canonical antibodies arise in sharks and camels? *PLoS Biol.* **2001**, *9*, e1001120.
5. Muyldermans, S. Nanobodies: Natural Single-Domain Antibodies. *Ann. Rev. Biochem.* **2013**, *82*, 775–797.
6. Zou, X.; Osborn, M.J.; Bolland, D.J.; Smith, J.A.; Corcos, D.; Hamon, M.; Oxley, D.; Hutchings, A.; Morgan, G.; Santos, F.; et al. Heavy chain-only antibodies are spontaneously produced in light chain-deficient mice. *J. Exp. Med.* **2007**, *204*, 3271–3283. [PubMed]
7. Janssens, R.; Dekker, S.; Hendriks, R.W.; Panayotou, G.; Van Remoortere, A.; Kong-a San, J.; Grosveld, F.; Drabek, D. Generation of heavy-chain-only antibodies in mice. *Proc. Natl.Acad. Sci. USA* **2006**, *103*, 15130–15135. [PubMed]
8. Nguyen, V.K.; Zou, X.; Lauwereys, M.; Brys, L.; Brüggemann, M.; Muyldermans, S. Heavy-chain only antibodies derived from dromedary are secreted and displayed by mouse B cells. *Immunology* **2003**, *109*, 93–101. [PubMed]
9. Hendershot, L.; Bole, D.; Köhler, G.; Kearney, J.F. Assembly and secretion of heavy chains that do not associate posttranslationally with immunoglobulin heavy chain-binding protein. *J. Cell Biol.* **1987**, *104*, 761–767. [PubMed]
10. Feige, M.J.; Groscurth, S.; Marcinowski, M.; Shimizu, Y.; Kessler, H.; Hendershot, L.M.; Buchner, J. An unfolded CH1 domain controls the assembly and secretion of IgG antibodies. *Mol. Cell.* **2009**, *34*, 569–579.
11. Leitzgen, K.; Knittler, M.R.; Haas, I.G. Assembly of immunoglobulin light chains as a prerequisite for secretion. A model for oligomerization-dependent subunit folding. *J. Biol. Chem.* **1997**, *272*, 3117–3123. [PubMed]
12. Kirkpatrick, R.B.; Ganguly, S.; Angelichio, M.; Griego, S.; Shatzman, A.; Silverman, C.; Rosenberg, M. Heavy Chain Dimers as Well as Complete Antibodies Are Efficiently Formed and Secreted from Drosophila via a BiP-mediated Pathway. *J. Biol. Chem.* **1995**, *270*, 19800–19805. [PubMed]
13. Haas, I.G.; Wabl, M.R. Immunoglobulin heavy chain toxicity in plasma cells is neutralized by fusion to pre-B cells. *Proc. Natl. Acad. Sci. USA* **1984**, *81*, 7185–7188. [PubMed]
14. Stoyle, C.L.; Stephens, P.E.; Humphreys, D.P.; Heywood, S.; Cain, K.; Bulleid, N.J. IgG light chain-independent secretion of heavy chain dimers: Consequence for therapeutic antibody production and design. *Biochem. J.* **2017**, *474*, 3179–3188.
15. Borth, N.; Mattanovich, D.; Kunert, R.; Katinger, H. Effect of increased expression of protein disulfide isomerase and heavy chain binding protein on antibody secretion in a recombinant CHO cell line. *Biotechnol. Prog.* **2005**, *21*, 106–111.
16. Kabsch, W. Integration, scaling, space-group assignment and postrefinement. *Acta Crystallogr. D Biol. Crystallogr.* **2010**, *66*, 133–144.
17. Vonrhein, C.; Flensburg, C.; Keller, P.; Sharff, A.; Smart, O.; Paciorek, W.; Womack, T.; Bricogne, G. Data processing and analysis with the autoPROC toolbox. *Acta Crystallogr. D Biol. Crystallogr.* **2011**, *67*, 293–302.
18. Winn, M.D.; Ballard, C.C.; Cowtan, K.D.; Dodson, E.J.; Emsley, P.; Evans, P.R.; Keegan, R.M.; Krissinel, E.B.; Leslie, A.G.; McCoy, A.; et al. Overview of the CCP4 suite and current developments. *Acta Crystallogr. D Biol. Crystallogr.* **2011**, *67*, 235–242.
19. McCoy, A.J.; Grosse-Kunstleve, R.W.; Adams, P.D.; Winn, M.D.; Storoni, L.C.; Read, R.J. Phaser crystallographic software. *J. Appl. Crystallogr.* **2007**, *40*, 658–674.
20. Terwilliger, T.C.; Grosse-Kunstleve, R.W.; Afonine, P.V.; Moriarty, N.W.; Zwart, P.H.; Hung, L.W.; Read, R.J.; Adams, P.D. Iterative model building, structure refinement and density modification with the PHENIX AutoBuild wizard. *Acta Crystallogr. D Biol. Crystallogr.* **2008**, *64*, 61–69.
21. Smart, O.S.; Womack, T.O.; Flensburg, C.; Keller, P.; Paciorek, W.; Sharff, A.; Vonrhein, C.; Bricogne, G. Exploiting structure similarity in refinement: Automated NCS and target-structure restraints in BUSTER. *Acta Crystallogr. D Biol. Crystallogr.* **2012**, *68*, 368–380. [PubMed]
22. Afonine, P.V.; Grosse-Kunstleve, R.W.; Echols, N.; Headd, J.J.; Moriarty, N.W.; Mustyakimov, M.; Terwilliger, T.C.; Urzhumtsev, A.; Zwart, P.H.; Adams, P.D. Towards automated crystallographic structure refinement with phenix.refine. *Acta Crystallogr. D Biol. Crystallogr.* **2012**, *68*, 352–367. [PubMed]
23. Emsley, P.; Cowtan, K. Coot: Model-building tools for molecular graphics. *Acta Crystallogr. D Biol. Crystallogr.* **2004**, *60*, 2126–2132. [PubMed]

24. Angal, S.; King, D.J.; Bodmer, M.W.; Turner, A.; Lawson, A.D.G.; Roberts, G.; Pedley, B.; Adair, J.R. A single amino acid substitution abolishes the heterogeneity of chimeric mouse/human (IgG4) antibody. *Mol. Immunol.* **1993**, *30*, 105–108.
25. Jayaram, N.; Bhowmick, P.; Martin, A.C.R. Germline VH/VL pairing in antibodies. *Protein Eng. Des. Sel.* **2012**, *25*, 523–530.
26. Shire, S.J. Formulation and manufacturability of biologics. *Curr. Opin. Biotechnol.* **2009**, *20*, 708–714.
27. Liu, H.; May, K. Disulfide bond structures of IgG molecules. *mAbs* **2012**, *4*, 17–23.
28. Froning, K.J.; Leaver-Fay, A.; Wu, X.; Phan, S.; Gao, L.; Huang, F.; Pustilnik, A.; Bacica, M.; Houlihan, K.; Chai, Q.; et al. Computational design of a specific heavy chain/kappa light chain interface for expressing fully IgG bispecific antibodies. *Protein Sci.* **2017**, *26*, 2021–2038.
29. Bailly, M.; Mieczkowski, C.; Juan, V.; Metwally, E.; Tomazela, D.; Baker, J.; Uchida, M.; Kofman, E.; Raoufi, F.; Motlagh, S.; et al. Predicting Antibody Developability Profiles Through Early Stage Discovery Screening. *mAbs* **2020**, *12*, 1–28.
30. Raybould, M.; Marksa, C.; Krawczyka, K.; Taddeseb, B.; Nowaka, J.; Lewisc, A.P.; Bujotzekd, A.; Shie, J.; Deanea, C.M. Five computational developability guidelines for therapeutic antibody profiling. *Proc. Natl. Acad. Sci. USA* **2019**, *116*, 4025–4030.
31. Haas, I.G.; Wabl, M.R. Immunoglobulin heavy chain binding protein. *Nature* **1983**, *306*, 387–389.
32. Feige, M.J.; Hendershot, L.M.; Buchner, J. How antibodies fold. *Trends Biochem. Sci.* **2010**, *35*, 189–198. [PubMed]
33. Wang, W.; Singh, S.; Zeng, D.; King, K.; Nema, S. Antibody Structure, Instability, and Formulation. *J. Pharm. Sci.* **2007**, *96*, 1–26. [PubMed]
34. Zhang, T.; Zhang, J.; Hewitt, D.; Tran, B.; Gao, W.; Qiu, Z.; Tejada, M.; Gazzano-Santoro, H.; Kao, Y.H. Identification and Characterization of Buried Unpaired Cysteines in a Recombinant Monoclonal IgG1 Antibody. *Anal. Chem.* **2012**, *84*, 7112–7123. [PubMed]
35. Lacy, E.R.; Baker, M.; Brigham-Burke, M. Free sulfhydryl measurement as an indicator of antibody stability. *Anal. Biochem.* **2008**, *382*, 66–68.
36. Franey, H.; Brych, S.R.; Kolvenbach, C.G.; Rajan, R.S. Increased aggregation propensity of IgG2 subclass over IgG1: Role of conformational changes and covalent character in isolated aggregates. *Protein Sci.* **2010**, *19*, 1601–1615.
37. Harris, R.J. Heterogeneity of recombinant antibodies: Linking structure to function. *Dev. Biol.* **2005**, *122*, 117–127.
38. Elkabetz, Y.; Argon, Y.; Bar-Nun, S. Cysteines in CH1 Underlie Retention of Unassembled Ig Heavy Chains. *J. Biol. Chem.* **2005**, *280*, 14402–14412.
39. Cherf, G.M.; Cochran, J.R. Applications of yeast surface display for protein engineering. *Methods Mol Biol.* **2015**, *1319*, 155–175.
40. Spiess, C.; Zhai, Q.; Carter, P.J. Alternative molecular formats and therapeutic applications for bispecific antibodies. *Mol. Immunol.* **2015**, *67*, 95–106.

*antibodies*

MDPI

*Article*

# Bringing the Heavy Chain to Light: Creating a Symmetric, Bivalent IgG-Like Bispecific

**Anusuya Ramasubramanian [1,*,†], Rachel Tennyson [1,†], Maureen Magnay [1,†,‡], Sagar Kathuria [1], Tara Travaline [1,§], Annu Jain [1,||], Dana M. Lord [1,¶], Megan Salemi [1], Caitlin Sullivan [1,**], Tristan Magnay [1], Jiali Hu [1], Eva Bric-Furlong [1], Pierrick Rival [2], Yanfeng Zhou [1], Dietmar Hoffmann [1], William Brondyk [1,††], Katarina Radošević [2,‡‡] and Partha S. Chowdhury [1,*,‡‡‡]**

1 Biologics Research, Sanofi R&D, 49 New York Avenue, Framingham, MA 01701, USA; rachel.tennyson@sanofi.com (R.T.); maureen.magnay@takeda.com (M.M.); sagar.kathuria@sanofi.com (S.K.); tara.travaline@gmail.com (T.T.); Annu.jain@cytiva.com (A.J.); dana.marie.lord@gmail.com (D.M.L.); Megan.Salemi@sanofi.com (M.S.); sullivcm@my.uri.edu (C.S.); tristan.magnay@sanofi.com (T.M.); jiali.hu@sanofi.com (J.H.); Eva.Bric-Furlong@sanofi.com (E.B.-F.); yanfeng.zhou@sanofi.com (Y.Z.); dietmar.hoffmann@sanofi.com (D.H.); whbrondyk@gmail.com (W.B.)

2 Biologics Research, Sanofi R&D, 13 Quai Jules Guesde, 94403 Vitry-sur-Seine, France; pierrick.rival@sanofi.com (P.R.); katarina.radosevic@lonza.com (K.R.)

* Correspondence: anusuya.ramasubramanian@sanofi.com (A.R.); pchowd15@its.jnj.com (P.S.C.)

† These authors contributed equally to this work.

‡ Current Address: Takeda, Cambridge, MA 02139, USA.

§ Current Address: FogPharma, Cambridge, MA 02140, USA.

|| Current Address: Cytiva, Marlborough, MA 01752, USA.

¶ Current Address: Ohana Biosciences, Inc., Cambridge, MA 02139, USA.

** Current Address: Inozyme Pharma, Inc., Boston, MA 02115, USA.

†† Current Address: Invetx, Boston, MA 02108, USA.

‡‡ Current Address: Lonza, 63360 Saint Beauzire, France.

‡‡‡ Current Address: Johnson & Johnson Pharmaceutical R&D, Spring House, PA 19477, USA.

Received: 21 August 2020; Accepted: 2 November 2020; Published: 6 November 2020

**Abstract:** Bispecific molecules are biologically significant, yet their complex structures pose important manufacturing and pharmacokinetic challenges. Nevertheless, owing to similarities with monoclonal antibodies (mAbs), IgG-like bispecifics conceptually align well with conventional expression and manufacturing platforms and often exhibit potentially favorable drug metabolism and pharmacokinetic (DMPK) properties. However, IgG-like bispecifics do not possess target bivalency and current designs often require tedious engineering and purification to ensure appropriate chain pairing. Here, we present a near-native IgG antibody format, the 2xVH, which can create bivalency for each target or epitope and requires no engineering for cognate chain pairing. In this modality, two different variable heavy (VH) domains with distinct binding specificities are grafted onto the first constant heavy (CH1) and constant light (CL) domains, conferring the molecule with dual specificity. To determine the versatility of this format, we characterized the expression, binding, and stability of several previously identified soluble human VH domains. By grafting these domains onto an IgG scaffold, we generated several prototype 2xVH IgG and Fab molecules that display similar properties to mAbs. These molecules avoided the post-expression purification necessary for engineered bispecifics while maintaining a capacity for simultaneous dual binding. Hence, the 2xVH format represents a bivalent, bispecific design that addresses limitations of manufacturing IgG-like bispecifics while promoting biologically-relevant dual target engagement.

**Keywords:** 2xVH; IgG-like bispecifics; bivalent bispecific molecule; symmetrical bispecific; concomittant binding; soluble VH

## 1. Introduction

The importance of antibodies as a therapeutic modality is now widely accepted as evidenced by the rapid rise of this drug class. Therapeutic monoclonal antibodies (mAbs), for instance, have maintained a dominant share of the drug market since 2017 and are expected to occupy nearly 20% of the drug market by 2022 [1,2]. Worldwide, over 570 mAbs have been studied in clinical trials and 97 have been approved in the US and Europe for a variety of indications [3,4]. Despite the commercial and clinical success of these molecules, the inherent monospecificity of mAbs is increasingly viewed as a bottleneck for their therapeutic potential. In particular, many diseases, including autoimmune disorders [5] and cancers [6,7], often have multiple immune modulators that cannot be effectively targeted by combinations of monospecific molecules and would therefore benefit from new mechanisms of actions associated with bi- and tri- specific molecules [8,9]. As such, the generation of bispecific antibodies, which take inspiration from natural monoclonal antibodies but are engineered for increased functionality, hold enormous clinical potential and address some of the limitations of therapeutic mAbs [9].

Currently, over one hundred bispecific formats have been described in the literature [10], each with its strengths and weaknesses. These different formats can be classified as either fragment-based bispecifics or IgG-based bispecifics [11]. However, due to the short half-life of fragment-based bispecifics, there has been an increased focus on designing IgG-based bispecifics [12,13]. This latter class of bispecifics, examples of which are shown in Figure 1a, benefit from the Fc and an associated increased half-life while also leveraging existing manufacturing and formulation infrastructure of IgGs [14–18]. Among these IgG-based bispecifics, there are some that use IgGs as scaffolds and graft either scFvs or Fabs to these scaffolds to create symmetric molecules [19,20] while others are IgG-like and create different specificities in the two Fab arms.

These asymmetrical IgG-like bispecifics possess potentially favorable drug metabolism and pharmacokinetic (DMPK) properties, owing to their similarities with traditional IgGs [17,18]. However, while these formats serve important niche applications, they suffer two important caveats: monovalent binding to each antigen and chain mispairing [11]. Since these IgG-like bispecifics bind each of their two targets in a monovalent fashion, they lose the target avidity of the monoclonal antibodies from which they are derived. Exceptions to this include the design proposed by Golay et al., which creates a tetravalent IgG-like bispecific using a CH1-CL heterodimerization strategy [19], or the Morrison type bispecifics that rely on grafting scFvs onto an IgG scaffold [21,22]. However, although these designs have been validated, they do not fully replicate the IgG structure. While there have been bispecific formats with a true IgG-like structure, such as the two-in-one antibody design [23] and the mAb$^2$ design [24], that aim to create bivalency for each specificity, it is yet to be demonstrated whether bivalent, bispecific interactions can be mediated by either format in a concomitant manner. As an additional consideration, IgG-like bispecific antibodies require extensive engineering to ensure cognate chain pairing between the heavy and light chains [11]. Even with techniques like "knob-into-holes" [25], strand-exchange engineered domains [26] and Fab interface engineering [27–29], these molecules continue to face important manufacturing hurdles relating to low yields or the removal of chain mispaired byproducts.

Here, we explore the possibility of generating a new bispecific design that aims to address the manufacturing and valency limitations of IgG-like bispecific molecules by leveraging the smallest immunoglobulin-based recognition unit, the VH domain. It has long been known that antigen-specific single human VH domains can exist in the absence of their cognate variable light domains (VLs) [30,31]. While stability of conventional IgGs depend on the VH/VL interaction, target-specific yet soluble and stable human or camelized human VH domains have been identified through both heavy-chain antibody (HcAb) transgenic mouse models [32–34] as well as phage display-based selections using semi-synthetic human VH libraries [35,36]. Recently, Shi et al. proposed a novel antibody format where two such soluble VH domains are used to replace the VH and VL of a traditional IgG through engraftment onto the $CH_1$ and CL domains of each Fab arm (Figure 1b) [37]. This proof of concept study used two anti-β-klotho VH domains to create biparatopic IgG-like molecules that show synergistic agonism unobserved in the individual soluble VH domains. Additionally, since these molecules were symmetrically constructed using autonomous VH domains, no special cognate chain pairing was required. However, the format's ability to tolerate individual VH domains targeting various antigens, a feature that would increase its broad, target-agnostic applicability, was unexplored.

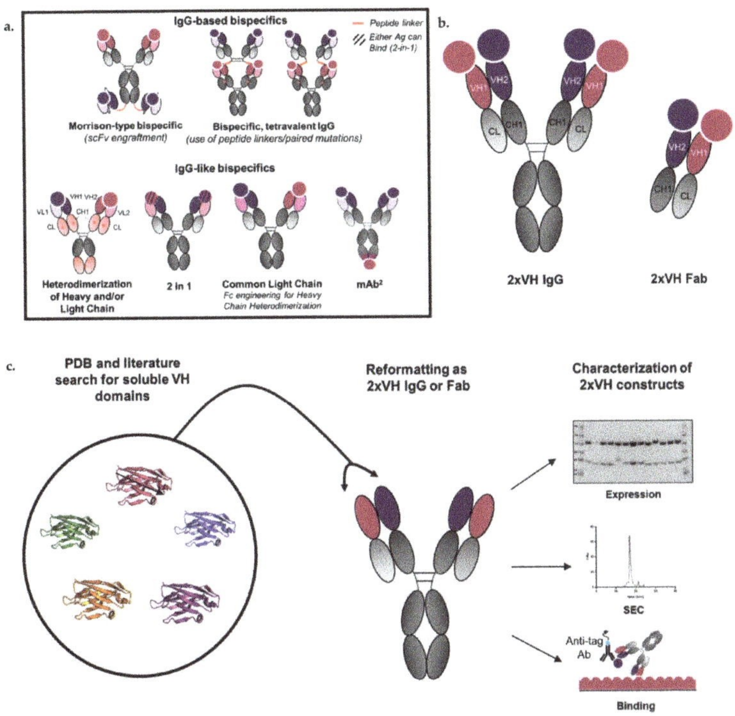


**Figure 1.** Examples of current bispecific IgG-based formats and overview of 2xVH IgG and Fab format. (**a**) Schematic depicting select examples of technologies used to generate IgG-based and IgG-like bispecifics including peptide linkers to generate tetravalent, bispecific IgGs, engraftment of scFvs on an IgG scaffold to generate Morrison-type bispecifics, heavy chain or light chain heterodimerization techniques, the 2-in-1 design, common light chain, and the utilization of Fcab via monoclonal antibodies ($mAb^2$). Red asterisks refer to the presence of mutations while red lines refer to peptide linkers. The capacity of each bispecific molecule to engage the two target antigens, depicted as circles, is shown. Hashed circles for the 2-in-1 format indicate that either antigen can engage with the antibody paratope. (**b**) Representation of proposed 2xVH IgG and Fab structure and antigen binding. (**c**) Workflow for the generation and characterization of 2xVH constructs derived from publicly available VH domains.

To assess the versatility of this format, referred to as 2xVH, we selected VH sequences with a range of sources and binding specificities from publicly available databases, including the protein databank (PDB). The criteria for selection included solubility following expression in cells, reasonable expression levels, a capacity for antigen binding and minimal surface hydrophobic patches as determined by BioVia Discovery Studio. Through this analysis, we selected thirteen previously identified VH domains. These domains were then characterized for expression and binding and further triaged before incorporation into the 2xVH IgG or Fab design. Once in the 2xVH format, the molecules were assessed for expression, purity and binding to their desired target antigens. Furthermore, we were able to validate dual binding in this format, including those targeting clinically relevant combinations, suggesting the modular nature of this format. Figure 1c schematizes this format modularity, particularly the capacity to integrate different VH domains identified from both in vitro and in vivo sources, as well as the initial steps for 2xVH molecule characterization. From our investigations, we believe the 2xVH format is a versatile bispecific, bivalent design that addresses some important limitations of generating IgG-like bispecifics [38] while showing promise as a dual- or multi-targeting therapeutic antibody format.

## 2. Materials and Methods

### 2.1. Selection of VH Domains from the Literature and the Protein Databank (PDB)

A review of both the protein databank (PDB) and the literature was performed to identify VH domains for production. The search included VH domains from in vitro display libraries [35,36] as well as in vivo sources including transgenic HcAb mice [32,34]. Relevant unique IDs from PDB were cross-referenced against the IMGT-3D structure-database for structural information [39], analyzed via BioVia Discovery Studio (Dassault Systèmes, Vélizy-Villacoublay, France) for hydrogen bonding, electrostatic interactions with its target antigen and surface hydrophobic patches. Structures were also annotated to identify chain and domain information. The VH domains from both PDB and literature searches were reviewed for their diversity of source (i.e., in vitro-derived or in vivo-derived) and the sequence availability to enable cloning. Domains were also ranked based on the commercial availability of target antigens. Additionally, available data from the literature regarding each domain's expression, solubility and capacity to mediate antigen binding were also considered. Based on these parameters and our outlined structural analysis, a set of thirteen VH domains were cloned, expressed and characterized for binding and expression as described below.

### 2.2. Expression and Purification of VH Domains

VH domains were cloned into a pHEN4 vector [40] using restriction enzymes NcoI and NotI (New England Biolabs, Ipswich, MA, USA). ExpiHEK293 cells were grown in 200 mL TB media (Sigma T-9179, Sigma-Aldrich, St. Louis, MO, USA) containing 0.8% glycerol, 2 mM $MgCl_2$, 0.1% glucose and 100 µg/mL ampicillin at 37 °C. Protein expression was induced with 1 mM IPTG when cells reached an optical density of 0.6–0.9 ($A_{600}$). Induction occurred at 28 °C, overnight, with shaking at 220 rpm. Cells were then pelleted and resuspended in 10 mL cold TES buffer. TES-suspended samples were placed on ice for 1 h. Simultaneously, TES buffer was diluted 4-fold in water and 15 mL of this diluted buffer was added to each sample and mixed via inversion. Samples were clarified via centrifugation at 5000× *g* for 30 min at 4 °C. Soluble lysate was allowed to bind 1 mL of Ni-NTA resin at 4 °C for 1 h while rotating. Resin was collected in disposable columns and washed with 5 column volumes (CV) of Buffer A (20 mM sodium phosphate, 500 mM NaCl, 5 mM imidazole pH 7.2). Proteins were then eluted with six CVs of Buffer B (20 mM sodium phosphate, 500 mM NaCl, 500 mM imidazole pH 7.2). Eluted pools were concentrated and buffer exchanged into Dulbecco's Phosphate Buffered Saline (DPBS) using 3K MWCO ultrafiltration membrane tubes (MilliporeSigma, Burlington, MA, USA). Analyses of total protein concentration and protein purity were performed using the Bradford assay and SDS-PAGE, respectively. Final protein concentrations were reported based on NanoDrop microvolume spectrophotometry.

### 2.3. Expression and Purification of 2xVH Fabs and IgG Molecules

For the generation of recombinant 2xVH Fab and IgG plasmids, human-codon optimized gBlock® DNA fragments were ordered from Integrated DNA Technologies (IDT) (Coralville, IA, USA). The gBlocks® were then cloned into a pTT5 mammalian expression vector using In-Fusion® Cloning technology (TaKaRa, Mountain View, CA, USA) to generate recombinant plasmids for heterologous expression in mammalian cells [41].

The fully automated mammalian cell secretory overexpression system, Protein Expression and Purification Platform (PEPP; GNF, San Diego, CA, USA), was used to express and to purify the bispecific antibody molecules. In brief, 35 mL of cultures seeded with $3.2 \times 10^6$ Expi293F cells (Thermo Fisher Scientific A14527, Waltham, MA, USA) were transfected with 34 μg of plasmid DNA using the Expifectamine 293 transfection reagent per the manufacturer's protocol (Life Technologies Corporation, Carlsbad, CA, USA). Cells were incubated at 37°C in an 8% $CO_2$ gas environment for 5 days. Culture supernatants were harvested and applied to MabSelect SuRe protein A resin (GE Healthcare 17-5438-02, Chicago, IL, USA) for gravity flow purification. Proteins were desalted using Nap-10 sephadex (GE Healthcare 17-0854-02) and eluted with 1X DPBS (Thermo Fisher Scientific 14190136), resulting in a final storage solution of phosphate buffered saline, pH 7.4 for each protein sample.

### 2.4. SPR Analysis

All surface plasmon resonance (SPR) analyses were done on a Biacore T100 using HBS-EP+ (GE Healthcare Life Sciences BR100669, Marlborough, MA, USA) running buffer. For dual binding experiments, 2xVH IgGs were diluted to 5 μg/mL in running buffer and captured on a Protein A Series S Sensor chip (GE Healthcare Life Sciences 29127555, Marlborough, MA, USA). The antigens were diluted to 100 nM in running buffer and each was sequentially flowed over the captured 2xVH IgG. For 2xVH Fabs, goat anti-human F(ab')2 (Jackson ImmunoResearch 109-005-006, West Grove, PA, USA) was amine coupled to a CM5 Series S sensor chip (GE Healthcare Life Sciences 29104988, Marlborough, MA, USA). The 2xVH Fabs were diluted to 20 μg/mL in running buffer and captured to the anti-F(ab')2 surface. The antigens were diluted to 100 nM in running buffer and each was sequentially flowed over the captured 2xVH IgG or Fab. All data were double referenced, and the curves were evaluated for binding responses above baseline.

For kinetics experiments, 2xVH IgGs were diluted to 5 μg/mL, 2xVH Fabs were diluted to 20 μg/mL and VH domains were diluted to 10 μg/mL in running buffer. The 2xVH IgGs were captured on a Protein A Series S sensor chip, 2xVH Fabs were captured on the CM5 Series S sensor chip immobilized with goat anti-human F(ab')2 and VH domains were captured on a CM5 Series S sensor chip immobilized with anti-his antibody (GE Healthcare Life Sciences 28995056, Marlborough, MA, USA). Antigens were diluted to 100 nM in running buffer and then serially diluted 2-fold for a total of six concentrations and a baseline, 0 nM control sample. Antigens were flowed over captured VHs from low to high concentration with a regeneration of 10 mM glycine-HCl pH 1.5 (GE Healthcare Life Sciences BR100354, Marlborough, MA, USA) in between each cycle. Resulting sensorgrams were double referenced and fit to a 1:1 binding model.

### 2.5. Analytical SEC

All size exclusion chromatography (SEC) analyses were done on an Agilent 1260 HPLC with a TSKgel G3000SWxl analytical column (Tosoh 0854, Tokyo, Japan) using phosphate buffered saline pH 6.8 as the mobile phase. For each sample, 20 μg of protein was injected in singlet over the column at 0.5 mL/min for 30 min and an absorbance at 280 nm (A280) was measured. Buffer blanks were included after each sample. Resulting chromatograms were blank subtracted and manually integrated to determine percent purity.

### 2.6. NanoDSF

Nano-format of Differential Scanning Fluorimetric (nanoDSF) was performed for purified VH domains using Prometheus NT (NanoTemper technologies PR001, Cambridge, MA, USA). Intrinsic tyrosine and tryptophan fluorescence were measured during thermal denaturation from room temperature to 95 °C at 1 °C/min ramp and 30% intensity.

## 2.7. ELISA

For dual-binding ELISAs, the surface immobilized antigen was coated overnight at 4 °C on an Immulon 4 HBX plate (Thermo Fisher Scientific 3855, Waltham, MA, USA) at 5 μg/mL. The plate was then washed three times with Phosphate Buffered Saline (PBS) solution using an automatic 96-well microplate washer (Titertek Berthold, Pforzheim, Germany) and blocked with 50 μL of 3% milk-PBS for 30 min. The blocked plates were subsequently washed twice with PBS supplemented with 0.1% Tween and once with PBS before varying concentrations of 2xVH IgG or Fab, diluted in 3% milk-PBS, were added to each well of the microtiter plate. The plates were incubated for one hour and subsequently washed twice with PBS supplemented with 0.1% Tween and once with PBS. The second tagged antigen, diluted to 5 μg/mL in 3% milk-PBS, was then added to the plate and incubated for 1 h before the plates were washed as previously described. Eu-labelled secondary antibody (Perkin Elmer, Waltham, MA, USA) targeting the tag (anti-Biotin, anti-His or anti-Fc) was then added at 0.2 μg/mL in DELFIA assay buffer (PerkinElmer, Waltham, MA, USA) and microtiter plates with secondary antibody were allowed to incubate in the dark for one hour. Plates were subsequently washed with DELFIA wash buffer and developed using the DELFIA enhancement solution according to the manufacturer's instructions. After 30 min of incubation, the fluorescence was measured via a microtiter plate reader at 615 nm.

## 2.8. 2xVH Fab Crystallization

Recombinant 2xVH anti-Lysozyme/Her2 Fab was expressed in Expi293F cells and purified by MabSelect SuRe protein A resin (GE Healthcare 17-5438-02, Chicago, IL, USA) followed by size exclusion chromatography (Superdex 200 10/300 GL-GE Healthcare; buffer: 20 mM HEPES pH 7.0, 50 mM NaCl). The protein was concentrated to 10 mg/mL and crystallized at room temperature in 1 M lithium chloride, 0.1 M citric acid pH 4.0. These crystals were cryoprotected in 20% ethylene glycol and mother liquor. X-ray diffraction data were collected at the Candian Macromolecular Crystallography Facility (CMCF) at the Canadian Light Source (CLS) using a Pilatus 3 6 M detector (Dectris, Philadelphia, PA, USA). The data were processed using XDS [42] and Aimless [43]. Molecular replacement was performed using Phaser and the data were refined using phenix.refine and the crystallographic object-oriented toolkit (COOT) [44–46]. Data collection and refinement statistics are listed (Table S1). Software used in this project was accessed through the SBGrid consortium [47]. Atomic coordinates have been deposited in the Protein Data Bank under accession code 7JKB.

# 3. Results

## 3.1. Identification of In Vitro and In Vivo-Derived Candidate VH Domains

To understand the capacity of the modular 2xVH bispecific IgG to tolerate diversely-sourced VH domains, we searched the protein databank (PDB) and the literature for soluble VH domains discovered through both in vitro and in vivo antibody discovery methods. Despite the generally poor biophysical properties of autonomous VH domains [32,34–36], our in silico search yielded thirteen soluble, human VH domains targeting commercially available antigens as seen in Figure 2a. Interestingly, eleven of the thirteen VH domains were members of the VH3 family—a master framework known to be more soluble and stable than any other human VH group in the absence of a cognate VL [48]. Despite this similarity, VH domains were derived from three distinct sources, namely phage-display libraries, transgenic mice or VH dimer (VHD) libraries, and some of the unique features of each class of domains are described below.

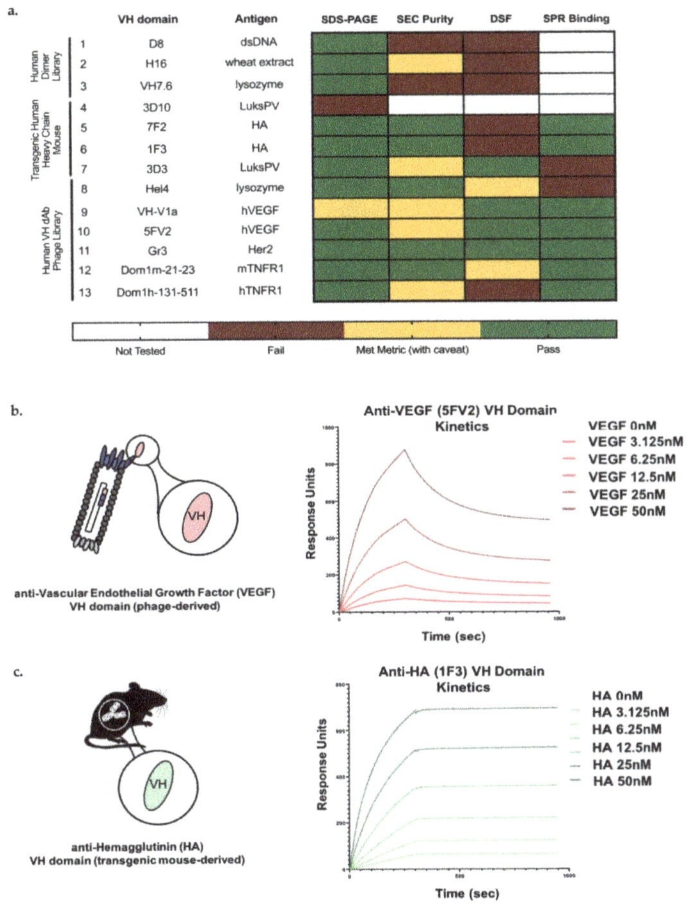


**Figure 2.** Shortlist of VH domains identified from literature and protein databank search. (**a**) A heatmap summarizing the thirteen VH domains from the literature chosen for further biophysical characterization. VH domains were characterized and triaged based on their expression, purity, melting temperature (*Tm*), and binding. Expression was assessed by recombinant protein production and visualization of soluble protein on an SDS-PAGE gel. Soluble protein was characterized as having passable expression (green), if the protein expressed at the correct molecular weight. VH domains that met this metric with caveat, shown in yellow, expressed but at an unexpected molecular weight, whereas those that failed (red) showed no protein expression. Likewise, for size exclusion chromatography (SEC), VH domains with a monomeric purity of 95% or higher are marked as green, while those that were 90–94.9% monomeric or ran at unexpected weight were marked in yellow. Any VH domain that was <90% monomer was flagged (red). Melting temperatures, $T_m$, were determined by nano-DSF and used as a measure of thermal stability. VH domains with a $T_m$ above 60 °C were the most thermostable (green). Any VH domains with a $T_m$ between 57.5–59.5 °C were considered moderately thermostable (yellow), while domains displaying $T_m$s below 57.5 °C had poor thermostability and are marked in red. Binding was analyzed by SPR for expressible VH domains and either marked in green for observable binding or in red for no observed binding. (**b**) Binding kinetics for phage-derived VH domain, 5FV2, binding to its target antigen, human vascular endothelial growth factor (VEGF). (**c**) Binding kinetics for transgenic mouse-derived 1F3 binding to its target antigen, hemagglutinin (HA). Both VH domains bind their target with affinities of 3 nM and 0.7 nM respectively.

In vitro sources, like phage display, can be used to identify non-aggregating VH domains against desired target antigens through panning synthetic or semi-synthetic libraries based on master frameworks like the V3-23 germline [49]. Analysis of published soluble VH (sVH) crystal structures in BioVia Discovery Studio suggested that incorporation of hydrophilic residues at the VL interface reduces hydrophobicity and can further improve solubility and any aggregation-associated issues that persist with these domains. In some cases, mutations of just the complementarity-determining regions or CDRs can be sufficient to induce a small conformational change that shields or alters a hydrophobic patch at the VH interface and thereby solubilizes the VH domain [50]. The solubilizing effect of mutations in the CDR was seen for one of our candidate VH domains, HEL4—a molecule that contains negatively charged residues in its CDR1 that causes a framework tryptophan to rotate inward. The rotation of this hydrophobic tryptophan side chain was proposed to have improved both the solubility and stability of HEL4 [51].

Although such solubilizing mutations can be rationally introduced to the CDRs of VH domains [52], phage display libraries also afford the opportunity to leverage stress-selections by heat [35], protease or acid treatment [53] to identify and produce stable VH domains. Previously, it was shown that three rounds of stress-selections on an error-prone PCR library derived from one of our candidate VH domains, Dom1h-131-511, could produce protease-optimized clones [54]. One such clone, DOM1h-131-203, showed a two-fold improvement in binding affinity for hTNFR1 and an increased melting temperature, a hallmark of improved thermostability among IgGs and VH domains [50,52]. As a consequence of this enormous flexibility afforded by phage display in tailoring library-based selections, numerous soluble VH (sVH) domains have been reported and six different phage-derived sVH domains were shortlisted for introduction in the 2xVH format—5FV2 targeting human VEGF [55], Dom1h-131-511 targeting human TNFR1 [54], Dom1m-21-23 targeting mouse TNFR1 [56], Gr3 targeting Her2 [57], Hel4 targeting lysozyme [35] and VH-V1a targeting human VEGF [58]. The criteria used for the selection of these VH domains as well as our in vivo-derived VH domains are outlined in the Materials and Methods.

In addition to in vitro discovery techniques, in vivo sources, like transgenic mice containing hybrid llama/human Ig heavy chains, can also produce HcAb upon immunization. Work by Drabek et al. showed that the VH portion of these HcAbs can be isolated and expressed as soluble proteins [59]. Upon sequence analysis, Drabek and colleagues found an increase in net hydrophobicity and the presence of charged residues in select CDR and framework regions, a feature that facilitates the solubility of these fully human VH domains. Consequently, four different transgenic mouse-derived VH domains bearing similar charged residues—1F3 and 7F2 targeting hemagglutinin [59], 3D3 and 3D10 targeting LuksPV [60]—were identified in silico and selected by us for further characterization and incorporation into the 2xVH format.

Aside from in vitro and in vivo discovery of native sVH domains, we were also interested in considering in vitro-derived VH dimers. It has been hypothesized that VH domains have the capacity to dimerize owing to the structural similarity between VH and VL and the capacity of the latter to form homodimers [61]. These VHDs create a VH:VH interface that is an appealing feature for the 2xVH format and similar in structure to a native VH:VL interface. Given this feature of VHDs, we decided to explore the capacity of VHD-derived VH domains to generate similar stabilizing interactions in the 2xVH context through three such domains: D8 targeting dsDNA [62], H16 targeting wheat extract [61] and VH7.6 targeting lysozyme [63]. Consequently, thirteen VH domains were incorporated into the 2xVH format, six of which were derived from the phage platform, four from transgenic mice and three from phage-derived VH dimer libraries.

### *3.2. Biophysical Characterization of sVH Candidates Shows Source-Dependent Differences in Expression, Purity and Thermal Stability*

We hypothesized that three key biophysical parameters—expression, purity and thermal stability—of soluble VH domains would have an important influence on the production of any derived 2xVH bispecific molecule. Therefore, each of the thirteen domain antibodies was expressed in a small-scale bacterial culture system, purified from the periplasm, and characterized via SDS-PAGE. Five of the six VH domains identified through phage-display uniformly expressed as soluble proteins by SDS-PAGE and all but one, VH-V1a being the exception, showed bands at their expected molecular weight. Of the four VH domains derived from transgenic mice, all but one, 3D10, could be expressed as a soluble protein. Similarly, all three VHD-derived VH domains showed uniform expression as soluble protein by SDS-PAGE. At this point, any VH domain that could not be easily produced in bacteria was triaged out, leaving twelve VH domains that were further characterized.

Measures of protein heterogeneity or purity of the monomeric fraction are essential for both antibodies and antibody fragments since the presence of impure species can result in aggregation [64] and other undesirable properties. To better understand the biophysical characteristics of these twelve soluble VH domains, we moved forward with subsequent analysis of purity by analytical size exclusion chromatography (aSEC) as shown in the classification scheme in Figure 2a. VH-V1a, despite having an SEC purity of 95.5%, was an exception to this classification scheme. Analysis by aSEC suggested the monomer was larger than anticipated, a finding that was further supported by SDS-PAGE analysis (data not shown). While the majority of phage- and transgenic-mouse derived VH domains showed a single monomeric peak in SEC chromatograms and purities above 90%, only one VHD-derived sVH (H16) showed comparable purity by aSEC (91.5% monomeric). By contrast, two VHD-derived VH domains, D8 and VH7.6, demonstrated a broad peak by aSEC, indicative of possible aggregates or unwanted isoforms of the VH domain [64]. In fact, these poor SEC profiles are corroborated by previous work illustrating the hydrophobicity of VHD dimerization interface [31]—a feature which, unless engineered or masked by pairing, could result in aggregation of the monomeric VH. Consequently, these molecules were flagged for triaging and are marked in 'red' in the heatmap in Figure 2a.

As with measures of protein aggregation, protein stability is a key developability parameter that can provide insight into the capacity of individual VH domains to maintain structural integrity as soluble domains and in the context of a bispecific format. Here, protein stability was addressed through thermal unfolding experiments where individual VH domains were subjected to a linear temperature ramp and thermal stability, quantified by the melting temperature ($T_m$), was monitored by nanoDSF. For the twelve identified VH domains the $T_m$s ranged from 47 °C to 65.4 °C, with six molecules showing higher $T_m$s above 58 °C. Unsurprisingly, VH domains identified from optimized phage display and transgenic mice showed higher $T_m$s (ranging from 55.9 to 65.2 °C and 54 to 57.1 °C, respectively), further supporting the idea that stress-selections and interface engineering play an important role in increasing the stability of these molecules. This is illustrated by DOM1h-131-511, a native VH domain targeting hTNFR1 that we initially shortlisted for incorporation in our bispecific 2xVH format. However, as suggested by Enever et al., this molecule showed a low $T_m$ (<58 °C) and monomeric purity (<90%) by aSEC (data not shown) [54]. Consequently, we instead decided to move forward with the protease-optimized variant, DOM1h-131-203, reported to have a higher $T_m$ of 64.4 °C by differential scanning calorimetry (DSC). We also observed that VHD-derived VH domains had the lowest $T_m$s at 47 and 53 °C, and these low melting temperatures, coupled with poor purity by SEC, eliminated these domains as candidates for the 2xVH format.

### 3.3. Candidate In Vitro- and In Vivo-Derived sVH Domains Show nM and Sub-nM Binding

Before generating bispecific molecules, we assessed if the individual sVH domains mediate strong, specific binding for their target antigen. To understand the kinetic parameters between each VH domain and corresponding target antigen, the nine shortlisted molecules were assessed by SPR and individually captured and oriented on the SPR chip surface before saturating concentrations of the cognate antigen were flowed over the chip. Two examples of VH kinetics, one for the phage-derived 5FV2 and another for the transgenic mouse-derived 1F3, are shown in Figure 2b,c. Fitting both these sensorgrams with a 1:1 binding model indicates high affinities of 3 nM for the 5FV2-hVEGF interaction and 0.7 nM for the 1F3–HA interaction. The high affinities can be attributed to 5FV2's fast on-rate

($7.3 \times 10^5$ $M^{-1}s^{-1}$) and 1F3's slow off-rate ($5.5 \times 10^{-5}$ $s^{-1}$). Figure S1 illustrates the binding interactions of the remaining seven VH domains showing measurable equilibrium dissociation constants ($K_D$) by SPR. The affinities of the different VHs for their cognate antigens ranged from 500 nM to 2.7 pM.

While these nine phage- and HcAb transgenic mouse-derived VH domains showed stable interactions with their target antigen, characterized by nM or sub-nM $K_D$s, two domains, 3D3 targeting Luks PV and HEL4 targeting lysozyme, were unable to bind their targets. Though the absence of target-specific binding was in contrast to the reported literature [34,35], it is possible that both VH domains require oriented antigen presentation, as suggested by Friguet et al. [65], or multivalency [35] to initiate binding to their target. While we did not assess the capacity to rescue binding through alternate antigen or VH presentations, both 3D3 and Hel4 displayed favorable purity and thermostability as monovalent sVH (Figure 2a). This suggests that elements of the 3D3 and Hel4 framework, derived from the VH3-23 germline, may confer stability and potentially serve as a scaffold for further engineering. For this reason, both 3D3 and Hel4 were pursued alongside the other seven VH domains for reformatting and assessment in the 2xVH format. Since two VH domains targeting VEGF and two targeting HA had been shortlisted, the anti-VEGF and anti-HA VH domains with lower SEC purity and $T_m$s, VH-V1a and 7F2, were triaged out. The seven, unique VH domains that ultimately progressed toward incorporation in the 2xVH scaffold were 1F3, 3D3, Hel4, 5FV2, Gr3, Dom1m-21-23 as well as the protease-optimized variant of Dom1h-131-511, Dom1h-131-203.

### 3.4. Generation and Characterization of 2xVH Molecules with High Yield and Purity

We were interested in understanding how each of these domains would behave when paired with another VH in the 2xVH format. To explore the degree of modularity in the 2xVH design, each of the seven characterized VH domains were combinatorially paired with the remaining six VH domains to generate twenty-one two-domain combinations. By accounting for two possible placements for each VH domain within the 2xVH format, forty-two unique 2xVH molecules were designed. Due to constraints in cloning and expression capacity, 31 IgGs and 21 Fabs were generated as illustrated in Figure 3.

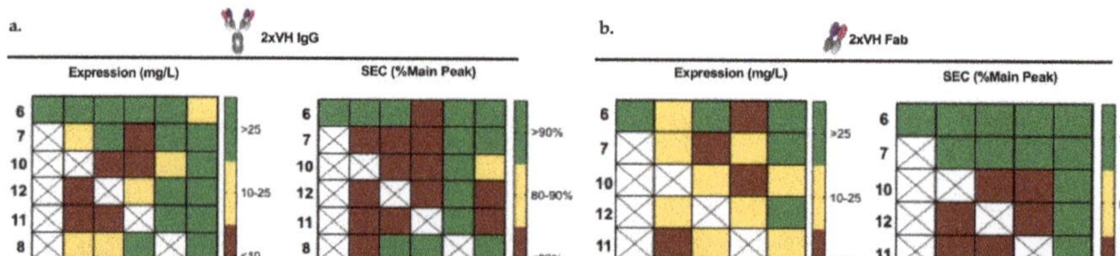


**Figure 3.** Characterization of 2xVH IgG and Fab expression and purity. (**a**) Heat map denoting the expression and purity of 2xVH IgG. $VH_1$, corresponding to the heavy chain position on the native IgG, is listed on the vertical axis of each heatmap while $VH_2$, corresponding to the light chain position, is listed on the horizontal axis of the heatmaps. The identity of the VH domain is delineated by a number scheme identical to the one outlined in Figure 2a. The expression, extrapolated into mg/L, is denoted in green for those IgGs expressing at greater than 25 mg/L, yellow for those showing moderate expression of 10–25 mg/L and red for any molecules with poor expression below 10 mg/L. SEC purity was calculated as a percentage of area under the main peak with 90% or higher monomeric purity marked in green, 80–90% monomeric purity marked in yellow and below 80% purity in red. Duplicate pairings (i.e., pairings with two identical VH domains) or combinations that were not produced are marked in white. (**b**) Heat map denoting expression and purity of corresponding 2xVH Fabs. Expression and SEC criteria were the same for both the 2xVH IgGs and Fabs.

Codon-optimized Fab- and IgG-versions of these molecules were cloned into a mammalian expression vector, expressed in small-scale cultures and assessed for expression and purity as shown by the heatmaps in Figure 3a,b, respectively. It is important to note that although our constituent sVHs were expressed and screened in *E. coli*, a flexible system for the expression of small or simple proteins, we chose to express the final 2xVH constructs in mammalian Expi293F cells since the expression of complex molecules like full-length IgGs is often difficult, if not impossible, in bacterial systems [66–68]. Overall, mammalian cell-expressed 2xVH Fabs showed similar expression with consistently higher purity compared to the corresponding 2xVH IgGs as seen in Figure 3b. Regardless of whether the VH pairs were expressed as Fabs or IgGs, certain $VH_1$–$VH_2$ combinations showed high expression and purity while other combinations showed poor tolerance for the bispecific format. For example, IgGs and Fabs with transgenic mouse-derived 1F3 in the $VH_1$ position or the phage-derived Hel4 and DOM1h-131-203 in the $VH_2$ position showed yields greater than 25 mg/L and over 90% purity. By contrast, other phage-derived VH domains including 5FV2 and Gr3, while stable as soluble VH fragments, showed poor expression and purity as 2xVH except when paired with 1F3, Hel4 or DOM1h-131-203.

To further investigate the stabilizing effects of these three VH domains, a small panel of molecules incorporating these domains—an anti-HA/anti-hVEGF (i.e., 1F3/5FV2), both orientations of the anti-hTNFR1/anti-hVEGF (i.e., 5FV2/DOM1h-131-203) and an anti-lysozyme/anti-Her2 (i.e., 5FV2/DOM1h-131-203)—were assessed for purity, binding and structure. The learnings from representative 2xVH constructs are discussed in the following sections.

### 3.5. 2xVH IgG and Fab Can Co-Engage Two Antigens: A Case Study of Anti-HA/Anti-hVEGF

To gain further understanding of the binding capacity of the 2xVH format and to ensure that VH domains originating from both in vivo and in vitro sources could integrate into a single 2xVH molecule, we generated prototype Fab and IgG molecules using a VH from transgenic mice, 1F3, and an optimized, phage-derived VH domain, 5FV2, as shown in Figure 4a. 1F3 was grafted onto CH1 of the heavy chain ($VH_1$ position) and 5FV2 was grafted onto the CL domain of the light chain ($VH_2$ position). Anti-HA/anti-hVEGF molecules expressed well as an IgG and showed a single peak by aSEC as seen in Figure 4b. While the Fab fragment expressed just shy of 25 mg/L, the molecule nevertheless eluted as a single peak (Figure 4c).

Next, to interrogate if these molecules could engage both their targets, HA and hVEGF, the 2xVH IgG and Fab molecules were assessed for binding by SPR. For this analysis, each 2xVH molecule was captured on an SPR sensor chip surface and sequentially exposed to each target antigen. We also assessed how dual binding differed if the molecule was first exposed to the 46 kDa hVEGF homodimer followed by the larger, 85 kDa HA as opposed to HA followed by hVEGF. While both the anti-HA/anti-hVEGF 2xVH IgG and the 2xVH Fab showed the capacity to bind the two target antigens, the binding profiles of the IgG and Fab differed significantly as seen in Figure 4d,e. In Figure 4d, the 2xVH IgG showed stable engagement and limited dissociation from both antigens regardless of the order in which the two targets were flowed over the IgG sample. By contrast, as shown in Figure 4e, hVEGF shows a partial, rapid dissociation from the 2xVH Fab both in the presence of existing HA interactions and in isolation. Furthermore, Fab engagement with HA was four-fold more attenuated when hVEGF interactions were already present than when HA was the first antigen to bind the 2xVH Fab. This suggested that while both 2xVH formats were able to engage with target antigens, the 2xVH IgG, by virtue of its bivalency, was able to cross-link VEGF homodimers and display stable antigen engagement. By contrast, the 2xVH Fab—a monovalent format whose binding capacity is more severely impacted by steric effects than the bivalent IgG—had more limited, sterically-driven target coengagement. Incidentally, this corroborates with the original characterization of 5FV2 performed by Walker et al. where a bivalent construct, consisting of two 5FV2 VH domains joined by a Gly-Ser linker, was shown to be more effective at binding and capturing the VEGF homodimer than a single, monovalent 5FV2 [55].

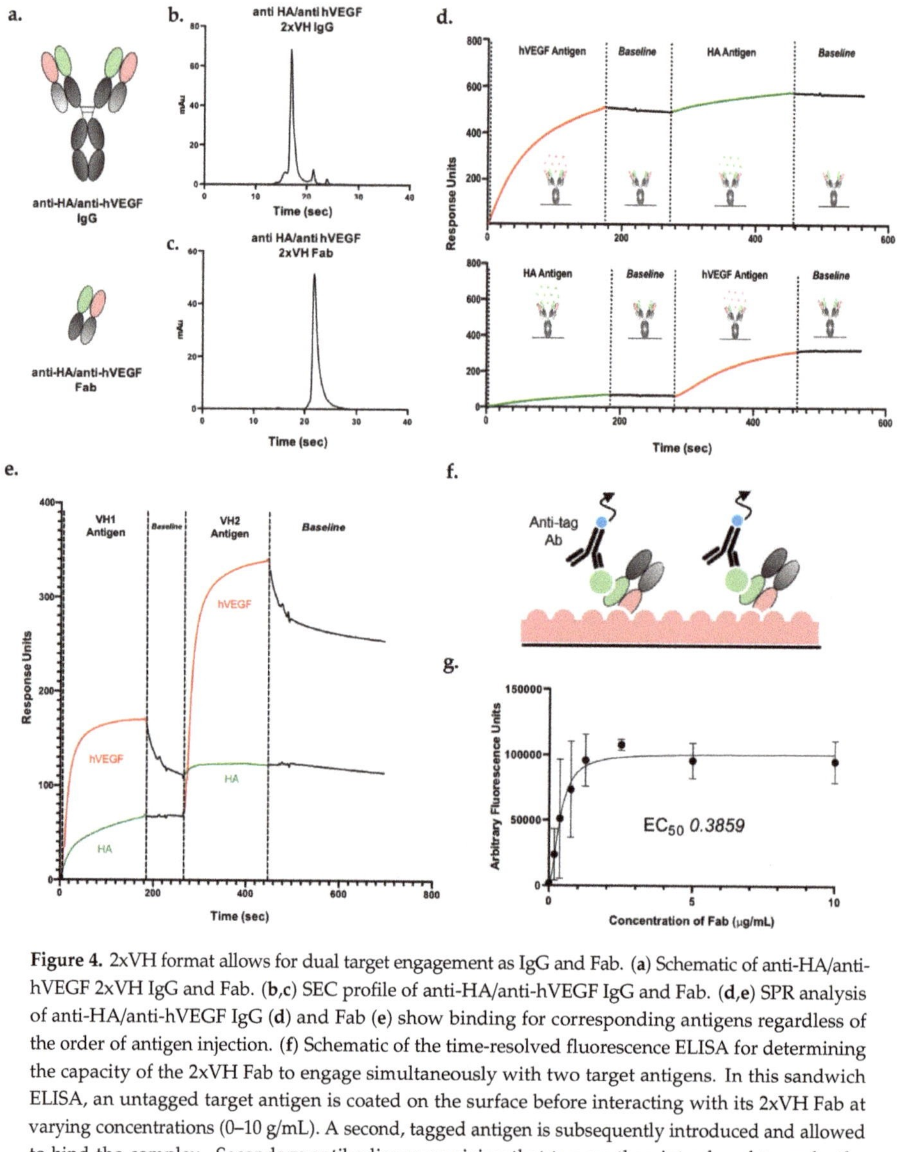


**Figure 4.** 2xVH format allows for dual target engagement as IgG and Fab. (**a**) Schematic of anti-HA/anti-hVEGF 2xVH IgG and Fab. (**b**,**c**) SEC profile of anti-HA/anti-hVEGF IgG and Fab. (**d**,**e**) SPR analysis of anti-HA/anti-hVEGF IgG (**d**) and Fab (**e**) show binding for corresponding antigens regardless of the order of antigen injection. (**f**) Schematic of the time-resolved fluorescence ELISA for determining the capacity of the 2xVH Fab to engage simultaneously with two target antigens. In this sandwich ELISA, an untagged target antigen is coated on the surface before interacting with its 2xVH Fab at varying concentrations (0–10 g/mL). A second, tagged antigen is subsequently introduced and allowed to bind the complex. Secondary antibodies recognizing that tag are then introduced to probe the antibody–dual antigen complex. A fluorescence signal at 655 nm occurs only if the Fab can bind both antigens at the same time. By contrast, the absence of a streptavidin-based signal suggests that both antigens cannot simultaneously bind the Fab, either because the Fab is unable to bind the surface-bound antigen (HA) or because second biotinylated antigen (hVEGF) is unable to bind to the Fab-HA complex. (**g**) ELISA results show concentration dependent binding of anti-HA/anti-hVEGF Fab to its corresponding antigens, HA and VEGF.

To explore if the 2xVH anti-HA/anti-hVEGF Fab could, in fact, simultaneously coengage with both hVEGF and HA, a modified sandwich ELISA assay was employed as shown in Figure 4f. In this scheme, the 2xVH Fab, at varying concentrations, was allowed to bind surface-adsorbed, untagged HA. Biotinylated hVEGF was subsequently allowed to interact with the Fab–HA complex and the entire trimeric complex was probed by a streptavidin-conjugated secondary antibody through a time-resolved fluorescence assay. Unlike with SPR, a positive signal is only observed in the sandwich ELISA when both HA and hVEGF bind to a single Fab. Using this fluorescence-based readout, Figure 4g illustrates that 2xVH anti-HA/anti-hVEGF Fab was able to mediate a dose-dependent, simultaneous co-engagement with both HA and hVEGF. Consequently, this binding data unambiguously suggest that both target antigens can simultaneously engage with their VHs within the same Fab arm of a 2xVH molecule—the first such illustration of simultaneous, dual binding for this format.

### *3.6. VH Domain Placement Does Not Impact Binding to Target Antigen in 2xVH Format: A Case Study of Clinically Relevant Anti-hVEGF/Anti-hTNFR1*

Having shown that prototype 2xVH molecules can engage in simultaneous, dual binding, we asked whether domain placement within the IgG could impact a molecule's capacity to bind its two, target antigens. To this end, two 2xVH IgGs incorporating 5FV2 targeting hVEGF, a mediator of angiogenesis in both development and disease, and DOM1h-131-203 targeting hTNFR1, the canonical receptor for the inflammatory TNF cytokine family, were generated and characterized. Notably, blocking both hVEGF and hTNF signaling has been observed to have therapeutic benefits in a number of autoimmune disorders including rheumatoid arthritis [69,70]. In the first molecule targeting these two antigens, the 5FV2 domain was grafted onto the CH1 domain or the $VH_1$ position while DOM1h-131-203 was grafted onto the CL, in the $VH_2$ position. In the inverted anti-hTNFR1/anti-hVEGF, the orientations were reversed and DOM1h-131-203 occupied the $VH_1$ position while the 5FV2 was grafted onto the $VH_2$ position. The schematics for these two molecules are shown in Figure 5a. Although 2xVH IgGs expressed well and their SEC profiles look to be 90% pure, we observed some minimal peak tailing suggesting column interactions and the possible presence of unresolved low molecular weight contaminants (Figure 5b,c). To address these concerns, SEC traces for both molecules were fit with exponentially modified Gaussian (EMG) curves as shown in Figure S2. While these fits suggest the possible presence of low-molecular weight contaminants, more than 70% of both samples are pure, monomeric IgG.

The initial binding assessment for these molecules was done by both SPR and ELISA. As illustrated in Figure 5d, both 2xVH IgGs show a similar capacity to engage in stable, dual binding by SPR when presented first with the $VH_1$ antigen followed by the $VH_2$ antigen. However, as with the prototype anti-HA/anti-hVEGF molecules, SPR-based analysis provides limited insight into whether the 5FV2- and DOM1h-131-203-containing bispecific can bind two antigens simultaneously or if these binding events are spatially and temporally distinct. To address this question and understand if these molecules engage antigens simultaneously, we employed a similar sandwich ELISA strategy used to characterize dual binding for the 2xVH anti-HA/anti-hVEGF Fab. In short, each 2xVH molecule could interact with either surface-immobilized hTNFR1 or hVEGF. The surface immobilized antigen-2xVH complex was then allowed to interact with the second tagged-antigen and probed with a secondary antibody targeting this tag. Figure 5e illustrates how each 2xVH IgG anti-hTNFR1/anti-hVEGF and anti-hVEGF/anti-hTNFR1 mediates dual binding when the $VH_1$ antigen is surface-immobilized and the $VH_2$ antigen is tagged, as well as, when the $VH_2$ antigen is surface-immobilized and the $VH_1$ antigen is tagged. While both 2xVH IgGs showed significant, dose-dependent dual binding, the measured $EC_{50}$s for these interactions varied depending on the presentation of the antigen. Namely, when the $VH_2$ antigen was surface-immobilized and initiated dual binding to the 2xVH molecule, the measured $EC_{50}$ for dual binding was ten-fold lower than when the $VH_1$ antigen was surface immobilized and the 2xVH molecule sequentially bound via $VH_1$ followed by $VH_2$. This suggests that this 2xVH molecule participates in a stronger dual binding response when the $VH_2$ antigen initiates dual binding. Conversely, when the $VH_1$ domain interacts with its target antigen first, the 2xVH molecule may have a weaker, subsequent interaction with the $VH_2$ antigen. While both 2xVH molecules mediated dual binding, the quality of this binding interaction can be influenced by sterics and vary based on both the domain placement and the order of antigen binding.

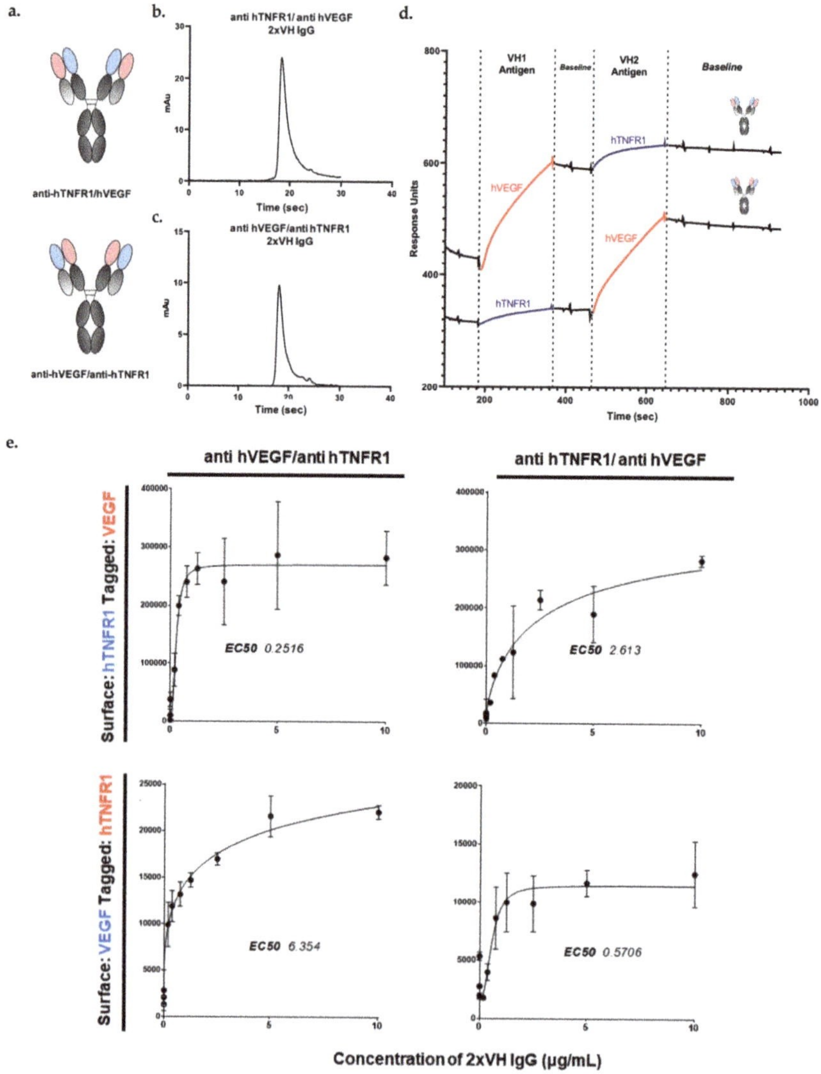


**Figure 5.** The 2xVH IgGs can undergo dual binding but domain placement and order of antigen binding can alter half-maximal effective concentrations or $EC_{50}$s (**a**) Schematic of anti-hTNFR1/anti-hVEGF and anti-hVEGF/anti-hTNFR1 2xVH IgGs. (**b**,**c**) Traces of monomeric purity for Anti-hTNFR1/anti-hVEGF IgG and hVEGF/anti-hTNFR1 IgG as assessed by analytical SEC. (**d**) SPR traces showing the sequential binding to the $VH_1$ and $VH_2$ antigen for the two mirror-image molecules, anti-hTNFR1/anti-hVEGF and anti-hVEGF/anti-hTNFR1 2xVH IgGs. (**e**) Time-resolved fluorescence ELISA results showing concentration dependent binding of the two mirror-image molecules, anti-hTNFR1/anti-hVEGF and anti-hVEGF/anti-hTNFR1. The top panels show dual binding of the two molecules when hTNFR1 is immobilized on the surface and hVEGF is tagged. In the bottom panels, the molecules are introduced to a hVEGF-immobilized surface and probed with tagged hTNFR1. While both 2xVH IgG can bind target antigens and display dual binding, domain placement and order of antigen binding can alter the $EC_{50}$ tenfold.

### 3.7. Anti-Her2/Anti-Lysozyme: VH/VH Interfaces Are Larger than Canonical VH/VL Interfaces

As mentioned previously, the heat maps in Figure 3a,b illustrate that the anti-lysozyme VH domain, Hel4, was one of the best tolerated VH domains in the 2xVH format, regardless of whether it occupied the $VH_1$ or $VH_2$ position within the molecule. To gain an understanding of why this VH domain was particularly amenable to the 2xVH format, we solved the crystal structure of the 2xVH anti-Her2/anti-lysozyme Fab schematized in Figure 6a, at a 2.55 Å resolution (Figure 6b).

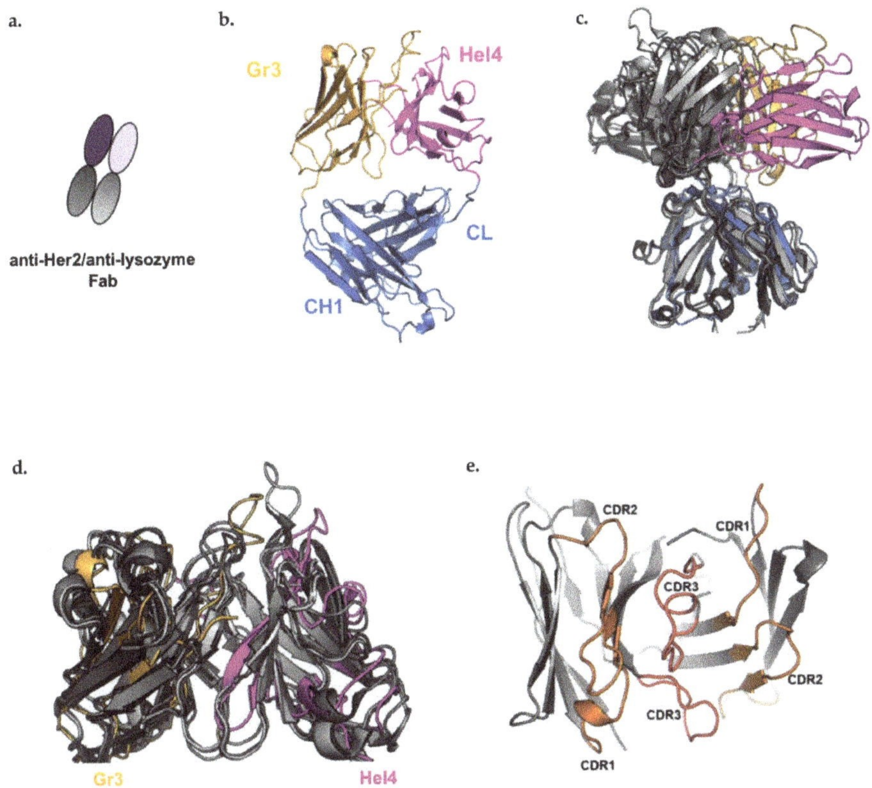


**Figure 6.** Crystal structure of 2xVH Fab shows VH:VH has a larger interface than natural VH:VL. (**a**) Schematic illustrating the configuration of VH domains in the 2xVH anti-Her2/anti-Lysozyme Fab. (**b**) A ribbon diagram illustrating the crystal structure of the 2xVH anti-Her2/anti-Lysozyme Fab, resolved to 2.55 Å. The Gr3 VH domain (shown in yellow) and the Hel4 domain (shown in magenta) were grafted to the CH1 and CL domains (shown in blue), respectively. (**c**) Superposition of the 2xVH (colored) with the following Fabs, derived from the VH 3-23 germline, in varying shades of gray: protein databank (PDB) 5I19, 5I1A, 5I1D, 5I1C. Structures are aligned on the CH1 domain. (**d**) Superposition of VH:VH domains from the 2xVH structure on canonical Fabs. (**e**) Top view of the VH:VH interface with CDR1 and 2 from both VH domains highlighted in orange and the long CDR3, thought to stabilize this interface, highlighted in red.

In order to explore the differences between the 2xVH structure and a VH/VL pair on a standard IgG, this 2xVH Fab was compared to the canonical pairings of VH 3-23 with various VLs shown in Table 1 [51]. Superposition of the 2xVH Fab structure on this conventional VH3-23:VL Fab shows the VH:VH orient differently from the VH:VL domains with respect to CH1 and CL. When the structures are superimposed on the CH1 domain, the CH1:CL domains of both the canonical Fab and the

2xVH Fab overlay well until the elbow region of the IgG. Upstream of the elbow region, the VH:VH domains of the 2xVH Fab do not align with the direction of VH:VL domains as shown in Figure 6c. However, the VH:VH dimer alone (in absence of CH1/CL) shows good alignment with VH:VL $F_v$ domains (root mean square deviation, R.M.S.D, of structures tested = ~1–2 Å), indicating that while the orientations of the domains differ, the structure of the domains themselves do not (Figure 6d).

**Table 1.** Buried Surface Area in Fab Structures.

| Construct | PDB | Contact Surface VH (Å$^2$) or *VH1* | Contact Surface VL (Å$^2$) or *VH2* | Interface (Å$^2$) |
|---|---|---|---|---|
| H3-23:L1-39 | 5I19 | 795 | 817 | 806 |
| H3-23:L3-11 | 5I1A | 822 | 834 | 828 |
| H3-23:L3-20 | 5I1C | 670 | 698 | 684 |
| H3-23:L4-1 | 5I1D | 743 | 770 | 757 |
| 2xVH anti-Her2, Lysozyme | 7JKB | 1200 | 1143 | 1171 |

To understand why this 2xVH Fab construct may have favorable stability, we looked at the interaction interface between the two VH domains via the buried surface area, commonly used to estimate the interface between two domains or macromolecules. The buried surface area was calculated using PISA (Protein Interfaces, Surfaces and Assemblies) on CCP4 by averaging the contact surface of each VH domain [71]. VH:VL contact surface area values for VH3-23:VL structures were extracted from Teplyakov et al. [72]. The buried surface area of VH:VH in the 2xVH anti-lysozyme/anti-Her2 Fab is significantly larger than the typical VH:VL contact surface in the referenced Fab structures [51] (Table 1). The VH:VH interface in the 2xVH anti-lysozyme/anti-Her2 Fab structure is 1171 Å$^2$, whereas the analogous VH:VL interfaces listed range from 684 to 828 Å$^2$. Many of the interactions at the 2xVH anti-lysozyme/Her2 Fab interface are contributed by the long, 16aa CDRH3 of the anti-Her2 VH, Gr3, as shown in Figure 6e. This CDR3 shares a long interaction interface with CDRH2 in Hel4 VH while also interacting with additional residues from CDRH1. This large VH–VH interface area, though unique and highly dependent on the length of the CDRH3, appears to be characteristic of the format and likely plays a role in stabilizing the 2xVH anti-lysozyme/anti-Her2 Fab and contributing to the overall stability of the 2xVH format.

## 4. Discussion

IgG-like bispecific antibodies, as well as their multispecific counterparts, have become a highly desirable modality for clinical therapeutics [8,9]. An attractive promise of this format is its capacity for dual targeting—a feature allowing for the creation of molecules that generate novel function by simultaneously binding and/or eliciting interaction between their targets. While there are over one hundred different bispecific formats that attest to the importance of this dual targeting capability [10], the majority of IgG-like bispecifics require considerable engineering to eliminate mispaired or unwanted species [11]. Additionally, IgG-like bispecifics, unlike their monospecific counterparts, bind each of their two targets in a monovalent fashion and thus do not display avidity-induced enhancements in activity. Although bispecific formats like the two-in-one antibody design aim to create bivalency for each specificity [23], it has yet to be demonstrated whether concomitant interactions with both targets occur on a single Fab arm. Here, we have presented and characterized a new format, termed the 2xVH bispecific, that overcomes many of the engineering and manufacturing hurdles currently plaguing IgG-like bispecifics and offers a simple, unique and modular format capable of bivalent targeting. Additionally, this design holds significant promise for generating multispecific molecules in the future.

The 2xVH format utilizes two soluble, fully human autonomous VH domains, each with its own target specificity, to replace the native VH and VL pair within a traditional IgG. Since binding to each target is determined by a single domain, it eliminates the need for cognate chain pairing. Moreover, the molecules' symmetry, with each half containing two specificities, gives rise to bivalency while retaining a nearly normal IgG structure. Consequently, these bispecific molecules can be produced via single-step purification methods and further polished for high purity via preparative SEC [11,37]. Previously, Shi and colleagues had developed a biparatopic 2xVH-like antibody targeting distinct and previously inaccessible epitopes on the β-Klotho/FGFR1c complex using two transgenic mouse-derived human sVH domains [37]. They showed that their biparatopic antibody was able to co-engage both epitopes of β-Klotho by biolayer interferometry (BLI) and inhibit downstream FGF21 signaling. However, BLI and other label-free optical analysis methods, which probe bulk changes in optical thickness at the biosensor tip, provide limited insight into whether dual binding events primarily occur on the same molecule at the same time and if so, whether these interactions are monovalent or bivalent in nature. Furthermore, the biparatopic anti-β-Klotho 2xVH molecules only make use of two VH domains, both derived from transgenic mice and targeting distinct epitopes on a single molecule [37]. Consequently, the capacity of this format to tolerate VH domains targeting two independent antigens was unexplored.

Given these limitations in the Shi et al. study, we searched the literature for publicly available VH domains that could be used to characterize and assess the versatility of the 2xVH format. After analyzing the expression, monomeric purity, thermal denaturation and binding properties of identified VH domains, we shortlisted seven different domains derived from both transgenic HcAb mice [32,34] and single domain VH phage display libraries [35,36]. Despite originating from different sources, all seven VH domains were derived from V3-23—a germline that is noted for its stability [49,51]. By grafting these seven VH domains into the 2xVH format in various configurations, we generated 31 unique 2xVH IgGs and 21 unique Fabs. While we observed that both phage display and transgenic mouse-derived VH domains could be substituted in the 2xVH format, the variability in expression and aSEC suggests that not every VH domain is well-tolerated. One of the best performing single VH domains, Gr3, showed greater than 90% monomeric purity by SEC, high $T_m$s indictive of thermostability and a sub-nM binding affinity for its target, Her2 (Figure 2a). Nevertheless, when Gr3 was incorporated into the $VH_1$ or $VH_2$ position of the 2xVH Fab and IgG formats in combination with a number of other VH partners, many of these Her2-targeting molecules were 10 to 20% monomeric, suggesting a propensity for aggregation.

By contrast, Hel4, a thermostable VH domain targeting lysozyme [51], was able to rescue this phenotype, with the 2xVH anti-lysozyme/anti-Her2 IgG showing a four-fold improvement in expression and 93% monomeric purity (Figure 3). Previous work had shown that HEL4 contains a unique feature—a rotated framework tryptophan—that improved both the solubility and stability of HEL4 VH domain [51] and may be implicated in the favorable pairing of HEL4 with the less stable Gr3. More specifically, with its framework tryptophan no longer surface exposed, HEL4 may display a more attractive interaction interface for other VH domains including Gr3. This observation suggests that although VH domains grafted onto the IgG scaffold do not chain pair as do traditional VH and VL domains, paired VH domains do interact, as shown in Figure 6, and the VH–VH interface plays an important stabilizing role for the molecule. In the future, we hope to generate crystal structures for a larger subset of these 2xVH molecules and better understand both the stability of the VH:VH interface as well as possible steric hindrance in the presence of antigen binding. Nevertheless, our analysis of the anti-lysozyme/anti-Her2 2xVH molecule suggests that many of its characteristics, particularly its monomeric purity and aggregation propensity, are emergent to the 2xVH format rather than intrinsic to the constituent VH domains. This characteristic affords the possibility of designing VH libraries to strategically select VH domains pairs or engineering the VH–VH interface to accommodate a wide range of VH domains, including those, like Gr3, that show poor biophysical properties as single sVH.

In addition to characterizing the 2xVH constructs by aSEC, we also assessed individual molecules' capacity to mediate binding with their target antigens as shown in Figures 4 and 5. SPR analysis of 2xVH IgG shows that these molecules are capable of dual engagement. However, as previously mentioned, label-free optical analysis methods provide limited information on whether dual targeting can occur simultaneously on the same molecule or whether each binding interaction is spatially and temporally distinct. To truly assess if the 2xVH format can co-engage its two target antigens on a single Fab arm, we performed a label-based sandwich ELISA utilizing a 2xVH Fab targeting HA and human VEGF. Within this scheme, a fluorescence-based signal, recognizing a biochemical tag of interest, is only observed if the entire binding complex of the immobilized antigen, the 2xVH Fab and the second antigen bearing the tag can form. With our 2xVH anti-HA/anti-hVEGF Fab, we observed a dose-dependent fluorescence signal indicating that a single Fab molecule was able to bind both a surface-immobilized HA and a tagged human VEGF. This work characterizing simultaneous dual antigen binding by Fab molecules is the first definitive evidence that two antigens can be bound by a Fab arm on a single 2xVH molecule. This also opens the possibility for leveraging this modality to generate multispecific molecules [11] in the future. However, based on this report, it should be noted that the 2xVH format does not ensure coengagement of two targets by the two VH domains within each Fab arm in every case. Rather, each new pair of VH domains within the 2xVH format must be tested for coengagement of cognate antigens.

Beyond definitively demonstrating dual engagement, our binding studies with the anti-hVEGF VH domain, 5FV2, also validate the early characterization of bivalent and monovalent versions of this domain by Walker et al. [55]. In their work, Walker and colleagues describe a bivalent, dual dAb generated by linking two 5FV2 VH domains utilizing a Gly-Ser linker. Notably, this dual dAb was more effective at binding and capturing the VEGF homodimer than a single 5FV2 domain, a feature the authors attributed to the dual dAb engaging in a "side-on" or "pincer"-like capture of the VEGF homodimer and the generation of a 2:2 heteromeric complex [55]. In our analysis we see a similar synergistic effect associated with bivalency, where the VEGF partially dissociates from the monovalent, bispecific 2xVH anti-HA/anti-hVEGF Fab but not from the bivalent, bispecific 2xVH anti-HA/anti-hVEGF IgG regardless of the antigen binding order (Figure 4d,e). This synergism suggests that the geometry, and more particularly the mutlivalency, of VH domains in the 2xVH IgG can aid in the targeting of structurally complex antigens [37] and antigen homodimers [55].

The role of geometry and the synergistic binding arising from bivalency was an important feature not only for the 2xVH anti-HA/anti-hVEGF IgG but also in the anti-hTNFR1 and anti-VEGF 2xVH IgG molecules. While these two VH domains were tolerated in both the $VH_1$ and $VH_2$ positions of the IgG scaffold, their positioning influenced their capacity to sequentially engage target antigens (Figure 5). In fact, we saw ten-fold higher $EC_{50}$ values, corresponding to a weaker dual target engagement, when the $VH_1$ domain bound its target antigen first as opposed to when the $VH_2$ domain initiated dual engagement. Though these observations are likely target dependent and were formulated based on a study of anti-hTNFR1 and anti-VEGF 2xVH IgG molecules, it has important implications for the design of clinically-relevant bispecific 2xVH where such sequential targeting may be required [12,73].

Our work highlights the potential versatility of a symmetric, simple yet novel bispecific format, the 2xVH. In addition to demonstrating that this format is amenable to two-chain expression with no engineering as well as single-step purification of both IgGs and Fabs, our work provides definitive evidence that a single 2xVH Fab can engage simultaneously with both its target antigens. This capacity for simultaneous dual engagement makes the 2xVH format a particularly attractive platform for the generation of bivalent, bispecific IgGs. Furthermore, when combined with technologies like "knob-into-hole" [25], orthogonal Fabs [55,74,75] or DuetMab [17], which increase the efficiency of heavy and light chain pairing, the bivalent, 2xVH technology affords the possibility of generating minimally engineered, asymmetric multispecifics including tri- or tetra-specific IgGs. In this vein, our work also illustrates the format's enormous flexibility, keeping in mind caveats associated with domain position and VH–VH pairing, for incorporating VH discovered through both in vitro and in vivo discovery methods. Given the widespread use of both single domain phage libraries [36,40,59] and transgenic heavy chain mice [32,34,59], we anticipate that target-specific VH domains can be rapidly discovered. By leveraging tailored selections and immunizations, we envision the generation of panels of VH domains with diverse affinity and epitope targeting capabilities and the strategic incorporation of these domains within the 2xVH scaffold to create functionally enhanced, clinically relevant multispecific therapeutics.

**Supplementary Materials:** The following are available online at http://www.mdpi.com/2073-4468/9/4/62/s1, Figure S1: SPR-Based binding profiles for six VH domains, Figure S2: Exponentially Modified Gaussian (EMG) fits for four lead 2xVH molecules (Fabs and IgGs), Table S1: Crystallography Data for the 2xVH anti-lysozyme/anti-Her2 Fab.

**Author Contributions:** The study was conceptualized by M.M., A.R., R.T. and P.S.C. with additional discussion of results and follow-up experimental strategy with K.R. and W.B. M.M. and P.R. performed the in silico analysis of VH domains. A.J., J.H., T.M., E.B.-F., M.M. and D.H. were involved in construct design and expression. S.K., C.S. and M.S. were involved in the biophysical characterization of all constructs. D.M.L. and Y.Z. were involved in the elucidation of 2xVH crystal structure. A.R., R.T. and T.T. performed all binding and kinetics experiments with the 2xVH molecules. Writing of the paper was done in close collaboration by A.R., R.T. and P.S.C. and technical feedback was provided by M.M., K.R., W.B., D.M.L. and Y.Z. All authors have read and agreed to the published version of the manuscript.

**Funding:** Sanofi provided support in the form of salaries to all the authors. This research received no external funding. The funders had no role in the design of the study; in the collection, analyses, or interpretation of data; in the writing of the manuscript, or in the decision to publish the results.

**Acknowledgments:** The authors would like to acknowledge Brian Mackness and Huawei Qiu for their technical support and invaluable feedback on experimental design, Catherine Prades and Catherine Devaud for their contributions to the PDB search and structure-based analysis of candidate VH domains, Samuel Stimple for his contributions to generating a study title and Ally Hatton for her business and legal support.

**Conflicts of Interest:** All authors are currently or were previously affiliated with Sanofi. The authors declare no conflict of interest.

## References

1. Tsumoto, K.; Isozaki, Y.; Yagami, H.; Tomita, M. Future perspectives of therapeutic monoclonal antibodies. *Immunotherapy* **2019**, *11*, 119–127. [CrossRef]
2. Akabane, H. Issues of the biopharmaceutical industry and recommendations for further development. *Ser. Res. Pap.* **2018**, *71*, 1–79.
3. Lu, R.M.; Hwang, Y.C.; Liu, I.J.; Lee, C.C.; Tsai, H.Z.; Li, H.J.; Wu, H.C. Development of therapeutic antibodies for the treatment of diseases. *J. Biomed. Sci.* **2020**, *27*, 1–30. [CrossRef] [PubMed]
4. Reichert, J.M. Antibody Therapeutics Approved or in Regulatory Review in the EU or US. Available online: https://www.antibodysociety.org/resources/approved-antibodies/ (accessed on 14 July 2020).
5. Schett, G.; Elewaut, D.; Mcinnes, I.B.; Dayer, J.; Neurath, M.F. How cytokine networks fuel inflammation. *Nat. Med.* **2013**, *19*, 822–826. [CrossRef]
6. Hanahan, D.; Weinberg, R. The Hallmarks of Cancer. *Cell* **2000**, *100*, 57–70. [CrossRef]
7. Faivre, S.; Djelloul, S.; Raymond, E. New Paradigms in Anticancer Therapy: Targeting Multiple Signaling Pathways With Kinase Inhibitors. *Semin. Oncol.* **2006**, *33*, 407–420. [CrossRef]
8. Zhao, Q. Bispecific Antibodies for Autoimmune and Inflammatory Diseases: Clinical Progress to Date. *BioDrugs* **2020**, *34*, 111–119. [CrossRef]
9. Husain, B.; Ellerman, D. Expanding the Boundaries of Biotherapeutics with Bispecific Antibodies. *BioDrugs* **2018**, *32*, 441–464. [CrossRef]
10. Brinkmann, U.; Kontermann, R.E. The making of bispecific antibodies. *mAbs* **2017**, *9*, 182–212. [CrossRef]
11. Krah, S.; Kolmar, H.; Becker, S.; Zielonka, S. Engineering IgG-Like Bispecific Antibodies—An Overview. *Antibodies* **2018**, *7*, 28. [CrossRef]
12. Labrijn, A.F.; Janmaat, M.L.; Reichert, J.M.; Parren, P.W.H.I. Bispecific antibodies: A mechanistic review of the pipeline. *Nat. Rev. Drug Discov.* **2019**, *18*, 585–608. [CrossRef]
13. Hladnik, L.; Augustin, K.; DeFrates, S. Advancements in Therapy for Acute Lymphoblastic Leukemia: Blinatumomab. *J. Adv. Pract. Oncol.* **2016**, *7*, 76–82. [CrossRef]
14. Merchant, A.M.; Zhu, Z.; Yuan, J.Q.; Goddard, A.; Adams, C.W.; Presta, L.G.; Carter, P. An efficient route to human bispecific IgG. *Nat. Biotechnol.* **1998**, *16*, 677–681. [CrossRef]
15. Jackman, J.; Chen, Y.; Huang, A.; Moffat, B.; Scheer, J.M.; Leong, S.R.; Lee, W.P.; Zhang, J.; Sharma, N.; Lu, Y.; et al. Development of a two-part strategy to identify a therapeutic human bispecific antibody that inhibits IgE receptor signaling. *J. Biol. Chem.* **2010**, *285*, 20850–20859. [CrossRef]

16. Tustian, A.D.; Endicott, C.; Adams, B.; Mattila, J.; Bak, H. Development of purification processes for fully human bispecific antibodies based upon modification of protein A binding avidity. *mAbs* **2016**, *8*, 828–838. [CrossRef] [PubMed]
17. Mazor, Y.; Oganesyan, V.; Yang, C.; Hansen, A.; Wang, J.; Liu, H.; Sachsenmeier, K.; Carlson, M.; Gadre, D.V.; Borrok, M.J.; et al. Improving target cell specificity using a novel monovalent bispecific IgG design. *mAbs* **2015**, *7*, 377–389. [CrossRef] [PubMed]
18. Chen, Y.; Xu, Y. Pharmacokinetics of Bispecific Antibody. *Curr. Pharmacol. Rep.* **2017**, *3*, 126–137. [CrossRef]
19. Golay, J.; Choblet, S.; Iwaszkiewicz, J.; Cérutti, P.; Ozil, A.; Loisel, S.; Pugnière, M.; Ubiali, G.; Zoete, V.; Michielin, O.; et al. Design and Validation of a Novel Generic Platform for the Production of Tetravalent IgG1-like Bispecific Antibodies. *J. Immunol.* **2016**, *196*, 3199–3211. [CrossRef]
20. Morrison, S.J.; Spradling, A.C.; Charnay, P.; Burns, D.K.; Parada, L.F.; Tong, W.G.; Ross, J.; Haug, J.; Johnson, T.; Feng, J.Q.; et al. Stem cells and niches: Mechanisms that promote stem cell maintenance throughout life. *Cell* **2008**, *132*, 598–611. [CrossRef]
21. Coloma, M.J.; Morrison, S.L. Design and production of novel tetravalent bispecific antibodies. *Nat. Biotechnol.* **1997**, *15*, 159–163. [CrossRef]
22. Marvin, J.S.; Zhu, Z. Recombinant approaches to IgG-like bispecific antibodies. *Acta Pharmacol. Sin.* **2005**, *26*, 649–658. [CrossRef]
23. Bostrom, J.; Yu, S.-F.; Kan, D.; Appleton, B.A.; Lee, C.V.; Billeci, K.; Man, W.; Peale, F.; Ross, S.; Wiesmann, C.; et al. Variants of the Antibody Herceptin That Interact with HER2 and VEGF at the Antigen Binding Site. *Science* **2009**, *323*, 1610–1614. [CrossRef]
24. Everett, K.L.; Kraman, M.; Wollerton, F.P.G.; Zimarino, C.; Kmiecik, K.; Gaspar, M.; Pechouckova, S.; Allen, N.L.; Doody, J.F.; Tuna, M. Generation of Fcabs targeting human and murine LAG-3 as building blocks for novel bispecific antibody therapeutics. *Methods* **2019**, *154*, 60–69. [CrossRef]
25. Ridgway, J.B.B.; Presta, L.G.; Carter, P. "Knobs-into-holes" engineering of antibody C(H)3 domains for heavy chain heterodimerization. *Protein Eng.* **1996**, *9*, 617–621. [CrossRef]
26. Davis, J.H.; Aperlo, C.; Li, Y.; Kurosawa, E.; Lan, Y.; Lo, K.M.; Huston, J.S. SEEDbodies: Fusion proteins based on strand-exchange engineered domain (SEED) CH3 heterodimers in an Fc analogue platform for asymmetric binders or immunofusions and bispecific antibodies. *Protein Eng. Des. Sel.* **2010**, *23*, 195–202. [CrossRef]
27. Bönisch, M.; Sellmann, C.; Maresch, D.; Halbig, C.; Becker, S.; Toleikis, L.; Hock, B.; Rüker, F. Novel CH1:CL interfaces that enhance correct light chain pairing in heterodimeric bispecific antibodies. *Protein Eng. Des. Sel.* **2017**, *30*, 685–696. [CrossRef]
28. Lewis, S.M.; Wu, X.; Pustilnik, A.; Sereno, A.; Huang, F.; Rick, H.L.; Guntas, G.; Leaver-Fay, A.; Smith, E.M.; Ho, C.; et al. Generation of bispecific IgG antibodies by structure-based design of an orthogonal Fab interface. *Nat. Biotechnol.* **2014**, *32*, 191–198. [CrossRef]
29. Liu, Z.; Leng, E.C.; Gunasekaran, K.; Pentony, M.; Shen, M.; Howard, M.; Stoops, J.; Manchulenko, K.; Razinkov, V.; Liu, H.; et al. A novel antibody engineering strategy for making monovalent bispecific heterodimeric IgG antibodies by electrostatic steering mechanism. *J. Biol. Chem.* **2015**, *290*, 7535–7562. [CrossRef]
30. Davies, J.; Riechmann, L. Antibody VH Domains as Small Recognition Units. *Bio/Technology* **1995**, *13*, 475–479. [CrossRef]
31. Hamers-Casterman, C.; Atarhouch, T.; Muyldermans, S.; Robinson, G.; Hammers, C.; Songa, E.B.; Bendahman, N.; Hammers, R. Naturally occurring antibodies devoid of light chains. *Nature* **1993**, *363*, 446–448. [CrossRef] [PubMed]
32. Janssens, R.; Dekker, S.; Hendriks, R.W.; Panayotou, G.; Van Remoortere, A.; San, J.K.A.; Grosveld, F.; Drabek, D. Generation of heavy-chain-only antibodies in mice. *Proc. Natl. Acad. Sci. USA* **2006**, *103*, 15130–15135. [CrossRef] [PubMed]
33. Craig, R.K.; Grosveld, F.G.; Janssens, R.W.; Drabek, D. Binding Molecules. WIPO Patent WO2006008548A3, 26 January 2006.
34. Laventie, B.J.; Rademaker, H.J.; Saleh, M.; De Boer, E.; Janssens, R.; Bourcier, T.; Subilia, A.; Marcellin, L.; Van Haperen, R.; Lebbink, J.H.G.; et al. Heavy chain-only antibodies and tetravalent bispecific antibody neutralizing Staphylococcus aureus leukotoxins. *Proc. Natl. Acad. Sci. USA* **2011**, *108*, 16404–16409. [CrossRef]

35. Jespers, L.; Schon, O.; Famm, K.; Winter, G. Aggregation-resistant domain antibodies selected on phage by heat denaturation. *Nat. Biotechnol.* **2004**, *22*, 1161–1165. [CrossRef]
36. Chen, W.; Zhu, Z.; Feng, Y.; Xiao, X.; Dimitrov, D.S. Construction of a Large Phage-Displayed Human Antibody Domain Library With a Scaffold Based on a Newly Identified Highly Soluble, Stable Heavy Chain Variable Domain. *J. Mol. Biol.* **2008**, *382*, 779–789. [CrossRef] [PubMed]
37. Shi, S.Y.; Lu, Y.W.; Liu, Z.; Stevens, J.; Murawsky, C.M.; Wilson, V.; Hu, Z.; Richards, W.G.; Michaels, M.L.; Zhang, J.; et al. A biparatopic agonistic antibody that mimics fibroblast growth factor 21 ligand activity. *J. Biol. Chem.* **2018**, *293*, 5909–5919. [CrossRef]
38. Carter, P.J.; Lazar, G.A. Next generation antibody drugs: Pursuit of the "high-hanging fruit". *Nat. Rev. Drug Discov.* **2018**, *17*, 197–223. [CrossRef]
39. Kaas, Q.; Ruiz, M.; Lefranc, M. IMGT/3Dstructure-DB and IMGT/StructuralQuery, a database and a tool for immunoglobulin, T cell receptor and MHC structural data. *Nucleic Acids Res.* **2004**, *32*, D208–D210. [CrossRef]
40. Arbabi Ghahroudi, M.; Desmyter, A.; Wyns, L.; Hamers, R.; Muyldermans, S. Selection and identification of single domain antibody fragments from camel heavy-chain antibodies. *FEBS Lett.* **1997**, *414*, 521–526. [CrossRef]
41. Durocher, Y. High-level and high-throughput recombinant protein production by transient transfection of suspension-growing human 293-EBNA1 cells. *Nucleic Acids Res.* **2002**, *30*, e9. [CrossRef]
42. Kabsch, W. XDS. *Acta Crystallogr. Sect. D Biol. Crystallogr.* **2010**, *66*, 125–132. [CrossRef]
43. Evans, P.R.; Murshudov, G.N. How good are my data and what is the resolution? *Acta Crystallogr. Sect. D Biol. Crystallogr.* **2013**, *69*, 1204–1214. [CrossRef]
44. McCoy, A.J.; Grosse-Kunstleve, R.W.; Adams, P.D.; Winn, M.D.; Storoni, L.C.; Read, R.J. Phaser crystallographic software. *J. Appl. Crystallogr.* **2007**, *40*, 658–674. [CrossRef]
45. Emsley, P.; Lohkamp, B.; Scott, W.G.; Cowtan, K. Features and development of Coot. *Acta Crystallogr. Sect. D Biol. Crystallogr.* **2010**, *66*, 486–501. [CrossRef] [PubMed]
46. Adams, P.D.; Afonine, P.V.; Bunkóczi, G.; Chen, V.B.; Davis, I.W.; Echols, N.; Headd, J.J.; Hung, L.W.; Kapral, G.J.; Grosse-Kunstleve, R.W.; et al. PHENIX: A comprehensive Python-based system for macromolecular structure solution. *Acta Crystallogr. Sect. D Biol. Crystallogr.* **2010**, *66*, 213–221. [CrossRef] [PubMed]
47. Morin, A.; Eisenbraun, B.; Key, J.; Sanschagrin, P.C.; Timony, M.A.; Ottaviano, M.; Sliz, P. Collaboration gets the most out of software. *eLife* **2013**, *2013*, e01456. [CrossRef]
48. Holt, L.J.; Herring, C.; Jespers, L.S.; Woolven, B.P.; Tomlinson, I.M. Domain antibodies: Proteins for therapy. *Trends Biotechnol.* **2003**, *21*, 484–490. [CrossRef]
49. Jirholt, P.; Ohlin, M.; Borrebaeck, C.A.K.; Söderlind, E. Exploiting sequence space: Shuffling in vivo formed complementarity determining regions into a master framework. *Gene* **1998**, *215*, 471–476. [CrossRef]
50. Nilvebrant, J.; Tessier, P.; Sidhu, S. Engineered Autonomous Human Variable Domains. *Curr. Pharm. Des.* **2016**, *22*, 6527–6537. [CrossRef] [PubMed]
51. Jespers, L.; Schon, O.; James, L.C.; Veprintsev, D.; Winter, G. Crystal structure of HEL4, a soluble, refoldable human VH single domain with a germ-line scaffold. *J. Mol. Biol.* **2004**, *337*, 893–903. [CrossRef]
52. Tiller, K.E.; Tessier, P.M. Advances in Antibody Design. *Annu. Rev. Biomed. Eng.* **2015**, *17*, 191–216. [CrossRef]
53. Famm, K.; Hansen, L.; Christ, D.; Winter, G. Thermodynamically Stable Aggregation-Resistant Antibody Domains through Directed Evolution. *J. Mol. Biol.* **2008**, *376*, 926–931. [CrossRef]
54. Enever, C.; Pupecka-Swider, M.; Sepp, A. Stress selections on domain antibodies: "What doesn't kill you makes you stronger". *Protein Eng. Des. Sel.* **2015**, *28*, 59–66. [CrossRef]
55. Walker, A.; Chung, C.W.; Neu, M.; Burman, M.; Batuwangala, T.; Jones, G.; Tang, C.M.; Steward, M.; Mullin, M.; Tournier, N.; et al. Novel interaction mechanism of a domain antibody-based inhibitor of human vascular endothelial growth factor with greater potency than ranibizumab and bevacizumab and improved capacity over aflibercept. *J. Biol. Chem.* **2016**, *291*, 5500–5511. [CrossRef]
56. Goodall, L.J.; Ovecka, M.; Rycroft, D.; Friel, S.L.; Sanderson, A.; Mistry, P.; Davies, M.L.; Stoop, A.A. Pharmacokinetic and pharmacodynamic characterisation of an anti-mouse TNF receptor 1 domain antibody formatted for in vivo half-life extension. *PLoS ONE* **2015**, *10*, e0137065. [CrossRef]

57. Baral, T.N.; Chao, S.Y.; Li, S.; Tanha, J.; Arbabi-Ghahroudi, M.; Zhang, J.; Wang, S. Crystal structure of a human single domain antibody dimer formed through V H-V H non-covalent interactions. *PLoS ONE* **2012**, *7*, e30149. [CrossRef]
58. Ma, X.; Barthelemy, P.A.; Rouge, L.; Wiesmann, C.; Sidhu, S.S. Design of synthetic autonomous VH domain libraries and structural analysis of a VH domain bound to vascular endothelial growth factor. *J. Mol. Biol.* **2013**, *425*, 2247–2259. [CrossRef]
59. Drabek, D.; Janssens, R.; de Boer, E.; Rademaker, R.; Kloess, J.; Skehel, J.; Grosveld, F. Expression cloning and production of human heavy-chain-only antibodies from murine transgenic plasma cells. *Front. Immunol.* **2016**, *7*, 3–12. [CrossRef]
60. Craig, R.K.; Grosveld, F.G.; Janssens, R.W.; Drabek, D.; Chen, T.; De Boer, E. Soluble Heavy-Chain Only Antibodies. U.S. Patent US 9365655 B2, 14 January 2016.
61. Sepúlveda, J.; Jin, H.; Sblattero, D.; Bradbury, A.; Burrone, O.R. Binders based on dimerised immunoglobulin VH domains. *J. Mol. Biol.* **2003**, *333*, 355–365. [CrossRef]
62. Jin, H.; Sepúlveda, J.; Burrone, O.R. Specific recognition of a dsDNA sequence motif by an immunoglobulin VH homodimer. *Protein Sci.* **2009**, *13*, 3222–3229. [CrossRef]
63. Jin, H.; Sepúlveda, J.; Burrone, O.R. Selection and characterisation of binders based on homodimerisation of immunoglobulin VH domains. *FEBS Lett.* **2003**, *554*, 323–329. [CrossRef]
64. Hong, P.; Koza, S.; Bouvier, E.S.P. A review size-exclusion chromatography for the analysis of protein biotherapeutics and their aggregates. *J. Liq. Chromatogr. Relat. Technol.* **2012**, *35*, 2923–2950. [CrossRef]
65. Friguet, B.; Chaffotte, A.F.; Djavadi-ohaniance, L.; Goldberg, M.E. Measurements of the true affinity constant in solution of antige-antibody complexes by ELISA—J. Immunol. Methods.pdf. *J. Immunol. Methods* **1985**, *77*, 305–319. [CrossRef]
66. Boss, M.A.; Kenten, J.H.; Wood, C.R.; Emtage, J.S. Assembly of functional antibodies from immunoglobulin heavy and light chains synthesised in *E. coli*. *Nucleic Acids Res.* **1984**, *12*, 3791–3806. [CrossRef]
67. Cabilly, S.; Riggs, A.D.; Pande, H.; Shively, J.E.; Holmes, W.E.; Rey, M.; Perry, L.J.; Wetzel, R.; Heyneker, H.L. Generation of antibody activity from immunoglobulin polypeptide chains produced in Escherichia coli. *Proc. Natl. Acad. Sci. USA* **1984**, *81*, 3273–3277. [CrossRef]
68. Hakim, R.; Benhar, I. "Inclonals": IgGs and IgG-enzyme fusion proteins produced in an *E. coli* expression-refolding system. *mAbs* **2009**, *1*, 281–287. [CrossRef]
69. Kim, H.R.; Kim, K.W.; Kim, B.M.; Cho, M.L.; Lee, S.H. The effect of vascular endothelial growth factor on osteoclastogenesis in rheumatoid arthritis. *PLoS ONE* **2015**, *10*, e0124909. [CrossRef]
70. Szekanecz, Z.; Koch, A. Targeting Angiogenesis in Rheumatoid Arthritis. *Curr. Rheumatol. Rev.* **2008**, *4*, 298–303. [CrossRef] [PubMed]
71. Krissinel, E.; Henrick, K. Inference of Macromolecular Assemblies from Crystalline State. *J. Mol. Biol.* **2007**, *372*, 774–797. [CrossRef]
72. Teplyakov, A.; Obmolova, G.; Malia, T.J.; Luo, J.; Muzammil, S.; Sweet, R.; Almagro, J.C.; Gilliland, G.L. Structural diversity in a human antibody germline library. *mAbs* **2016**, *8*, 1045–1063. [CrossRef]
73. Mullard, A. Bispecific antibody pipeline moves beyond oncology. *Nat. Rev. Drug Discov.* **2017**, *16*, 666–668. [CrossRef] [PubMed]
74. Leaver-Fay, A.; Froning, K.J.; Atwell, S.; Hector, A.; Pustilnik, A.; Lu, F.; Huang, F.; Yuan, R.; Hassanali, S.; Chamberlain, A.K.; et al. Computationally Designed Bispecific Antibodies Using Negative State Repertoires. *Structure* **2016**, *24*, 641–651. [CrossRef]
75. Froning, K.J.; Leaver-Fay, A.; Wu, X.; Phan, S.; Gao, L.; Huang, F.; Pustilnik, A.; Bacica, M.; Houlihan, K.; Chai, Q.; et al. Computational design of a specific heavy chain/κ light chain interface for expressing fully IgG bispecific antibodies. *Protein Sci.* **2017**, *26*, 2021–2038. [CrossRef] [PubMed]

MDPI
St. Alban-Anlage 66
4052 Basel
Switzerland
Tel. +41 61 683 77 34
Fax +41 61 302 89 18
www.mdpi.com

*Antibodies* Editorial Office
E-mail: antibodies@mdpi.com
www.mdpi.com/journal/antibodies

www.ingramcontent.com/pod-product-compliance
Lightning Source LLC
Chambersburg PA
CBHW062101220926
51850CB00010B/2847
* 9 7 8 3 0 3 6 5 7 6 9 0 9 *